D1691574

**Adaptive Finite Element
Solution Algorithm
for the Euler Equations**

by Richard A. Shapiro

Notes on Numerical Fluid Mechanics (NNFM) Volume 32

Series Editors: Ernst Heinrich Hirschel, München
Kozo Fujii, Tokyo
Bram van Leer, Ann Arbor
Keith William Morton, Oxford
Maurizio Pandolfi, Torino
Arthur Rizzi, Stockholm
Bernard Roux, Marseille

(Adresses of the Editors: see last page)

Volume 6	Numerical Methods in Laminar Flame Propagation (N. Peters / J. Warnatz, Eds.)
Volume 7	Proceedings of the Fifth GAMM-Conference on Numerical Methods in Fluid Mechanics (M. Pandolfi / R. Piva, Eds.)
Volume 8	Vectorization of Computer Programs with Applications of Computational Fluid Dynamics (W. Gentzsch)
Volume 9	Analysis of Laminar Flow over a Backward Facing Step (K. Morgan / J. Periaux / F. Thomasset, Eds.)
Volume 10	Efficient Solutions of Elliptic Systems (W. Hackbusch, Ed.)
Volume 11	Advances in Multi-Grid Methods (D. Braess / W. Hackbusch / U. Trottenberg, Eds.)
Volume 12	The Efficient Use of Vector Computers with Emphasis on Computational Fluid Dynamics (W. Schönauer / W. Gentzsch, Eds.)
Volume 13	Proceedings of the Sixth GAMM-Conference on Numerical Methods in Fluid Mechanics (D. Rues / W. Kordulla, Eds.) (out of print)
Volume 14	Finite Approximations in Fluid Mechanics (E. H. Hirschel, Ed.)
Volume 15	Direct and Large Eddy Simulation of Turbulence (U. Schumann / R. Friedrich, Eds.)
Volume 16	Numerical Techniques in Continuum Mechanics (W. Hackbusch / K. Witsch, Eds.)
Volume 17	Research in Numerical Fluid Dynamics (P. Wesseling, Ed.)
Volume 18	Numerical Simulation of Compressible Navier-Stokes Flows (M. O. Bristeau / R. Glowinski / J. Periaux / H. Viviand, Eds.)
Volume 19	Three-Dimensional Turbulent Boundary Layers – Calculations and Experiments (B. van den Berg / D. A. Humphreys / E. Krause / J. P. F. Lindhout)
Volume 20	Proceedings of the Seventh GAMM-Conference on Numerical Methods in Fluid Mechanics (M. Deville, Ed.)
Volume 21	Panel Methods in Fluid Mechanics with Emphasis on Aerodynamics (J. Ballmann / R. Eppler / W. Hackbusch, Eds.)
Volume 22	Numerical Simulation of the Transonic DFVLR-F5 Wing Experiment (W. Kordulla, Ed.)
Volume 23	Robust Multi-Grid Methods (W. Hackbusch, Ed.)
Volume 24	Nonlinear Hyperbolic Equations – Theory, Computation Methods, and Application (J. Ballmann / R. Jeltsch, Eds.)
Volume 25	Finite Approximation in Fluid Mechanics II (E. H. Hirschel, Ed.)
Volume 26	Numerical Solution of Compressible Euler Flows (A. Dervieux / B. van Leer / J. Periaux / A. Rizzi, Eds.)
Volume 27	Numerical Simulation of Oscillatory Convection in Low-Pr Fluids (B. Roux, Ed.)
Volume 28	Vortical Solutions of the Conical Euler Equations (K. G. Powell)
Volume 29	Proceedings of the Eighth GAMM-Conference on Numerical Methods in Fluid Mechanics (P. Wesseling, Ed.)
Volume 30	Numerical Treatment of the Navier-Stokes Equations (W. Hackbusch / R. Rannacher, Eds.)
Volume 31	Parallel Algorithms for Partial Differential Equations (W. Hackbusch, Ed.)
Volume 32	Adaptive Finite Element Solution Algorithm for the Euler Equations (R. A. Shapiro)

Adaptive Finite Element Solution Algorithm for the Euler Equations

by
Richard A. Shapiro

Die Deutsche Bibliothek - CIP-Einheitsaufnahme

Shapiro, Richard A.:
Adaptive finite element solution algorithm
for the Euler equations / by Richard A. Shapiro. -
Braunschweig: Vieweg, 1991
  (Notes on numerical fluid mechanics; Vol. 32)
  ISBN 3-528-07632-1
NE: GT

Manuscripts should have well over 100 pages. As they will be reproduced photomechanically they should be typed with utmost care on special stationary which will be supplied on request.
In print, the size will be reduced linearly to approximately 75 per cent. Figures and diagrams should be lettered accordingly so as to produce letters not smaller than 2 mm in print. The same is valid for handwritten formulae. Manuscripts (in English) or proposals should be sent to the general editor, Prof. Dr. E. H. Hirschel, Herzog-Heinrich-Weg 6, D-8011 Zorneding.

Vieweg is a subsidiary company of the Bertelsmann Publishing Group International.

All rights reserved
© Friedr. Vieweg & Sohn Verlagsgesellschaft mbH, Braunschweig 1991

No part of this publication may be reproduced, stored in a retrieval system or transmitted, mechanical, photocopying or otherwise, without prior permission of the copyright holder.

Produced by W. Langelüddecke, Braunschweig
Printed on acid-free paper
Printed in the Federal Republic of Germany

ISSN 0179-9614

ISBN 3-528-07632-1

Foreword

This monograph is the result of my PhD thesis work in Computational Fluid Dynamics at the Massachusettes Institute of Technology under the supervision of Professor Earll Murman. A new finite element algorithm is presented for solving the steady Euler equations describing the flow of an inviscid, compressible, ideal gas. This algorithm uses a finite element spatial discretization coupled with a Runge-Kutta time integration to relax to steady state. It is shown that other algorithms, such as finite difference and finite volume methods, can be derived using finite element principles. A higher-order biquadratic approximation is introduced. Several test problems are computed to verify the algorithms. Adaptive gridding in two and three dimensions using quadrilateral and hexahedral elements is developed and verified. Adaptation is shown to provide CPU savings of a factor of 2 to 16, and biquadratic elements are shown to provide potential savings of a factor of 2 to 6. An analysis of the dispersive properties of several discretization methods for the Euler equations is presented, and results allowing the prediction of dispersive errors are obtained. The adaptive algorithm is applied to the solution of several flows in scramjet inlets in two and three dimensions, demonstrating some of the varied physics associated with these flows. Some issues in the design and implementation of adaptive finite element algorithms on vector and parallel computers are discussed.

Many people have contributed to this research, and I am very grateful for all the help I have received. Professors Earll Murman, Mike Giles, Lloyd N. Trefethen, and Saul Abarbanel of M.I.T. have provided many insights into the fluid mechanical problems and the mathematics behind their solution. I also want to thank Dr. Rainald Löhner, Prof. Ken Morgan and Prof. Earl Thornton for all the ideas, conversations and criticisms they provided. My colleagues in the CFD lab have been wonderful foils for ideas and great helps in pointing out the obvious (and not-so-obvious) problems in this research effort. I thank my family for encouragement throughout my time at MIT. Most of all, I want to thank my wife Heather for helping me keep things in perspective and for all her support throughout my time as a graduate student. Her contributions to this thesis, direct and indirect, I will treasure always.

This work was supported in part by the Air Force Office of Scientific Research under contracts AFOSR-87-0218 and AFOSR-82-0136, and in part by the Fannie and John Hertz Foundation.

R. A. S.

—Proverbs 3:5,6

Contents

1 Introduction   1
    1.1 Research Goals 1
    1.2 Overview of Thesis 2
    1.3 Survey of Finite Element Methods for the Euler Equations   3

2 Governing Equations   5
    2.1 Euler Equations 5
    2.2 Non-Dimensionalization of the Equations 7
    2.3 Auxiliary Quantities 8
    2.4 Boundary Conditions 8
        2.4.1 Solid Surface Boundary Conditions 9
        2.4.2 Open Boundary Conditions 9

3 Finite Element Fundamentals   11
    3.1 Basic Definitions 11
    3.2 Finite Elements and Natural Coordinates 13
        3.2.1 Properties of Interpolation Functions 14
        3.2.2 Natural Coordinates and Derivative Calculation . 15

	3.3	Typical Elements	17
		3.3.1 Bilinear Element	17
		3.3.2 Biquadratic Element	19
		3.3.3 Trilinear Element	20
4	**Solution Algorithm**		**22**
	4.1	Overview of Algorithm	22
	4.2	Spatial Discretization	23
	4.3	Choice of Test Functions	25
		4.3.1 Test Functions for Galerkin Method	26
		4.3.2 Test Functions for Cell-Vertex Method	27
		4.3.3 Test Functions for Central Difference Method . . .	29
	4.4	Boundary Conditions	31
		4.4.1 Solid Surface Boundary Condition	31
		4.4.2 Open Boundary Condition	32
	4.5	Smoothing	34
		4.5.1 Conservative, Low-Accuracy Second Difference . .	34
		4.5.2 Non-Conservative, High-Accuracy Second Difference	35
		4.5.3 Combined Smoothing	37
		4.5.4 Smoothing on Biquadratic Elements	38
	4.6	Time Integration	38
	4.7	Consistency and Conservation	39
		4.7.1 Making Artificial Viscosity Conservative	40

5 Algorithm Verification and Comparisons 42

    5.1 Introduction 42

    5.2 Verification and Comparison of Methods 43

        5.2.1 5° Converging Channel 43

        5.2.2 15° Converging Channel 45

        5.2.3 4% Circular Arc Bump 46

        5.2.4 10% Circular Arc Bump 47

        5.2.5 10% Cosine Bump 47

        5.2.6 CPU Comparison and Recommendations 48

        5.2.7 Verification of Conservation 48

    5.3 Effects of Added Dissipation 49

    5.4 Biquadratic *vs.* Bilinear 51

        5.4.1 5° Channel Flow 51

        5.4.2 4% Circular Arc Bump 52

        5.4.3 10% Cosine Bump 52

    5.5 Three Dimensional Verification 53

    5.6 Summary 54

6 Adaptation 76

    6.1 Introduction 76

    6.2 Adaptation Procedure 77

        6.2.1 Placement of Boundary Nodes 79

        6.2.2 How Much Adaptation? 80

- 6.3 Adaptation Criteria 80
  - 6.3.1 First-Difference Indicator 81
  - 6.3.2 Second-Difference Indicator 82
  - 6.3.3 Two-Dimensional Directional Adaptation 83
- 6.4 Embedded Interface Treatment 84
  - 6.4.1 Two-Dimensional Interface 84
  - 6.4.2 Three-Dimensional Interface 87
- 6.5 Examples of Adaptation 88
  - 6.5.1 Multiple Shock Reflections 88
  - 6.5.2 4% Circular Arc Bump 89
  - 6.5.3 10% Circular Arc Bump 90
  - 6.5.4 3-D Channel 90
  - 6.5.5 Distorted Grid 91
- 6.6 CPU Time Comparisons 91

7 Dispersion Phenomena and the Euler Equations — 103

- 7.1 Introduction 103
- 7.2 Difference Stencils 103
  - 7.2.1 Some Properties of the Galerkin Stencil 104
  - 7.2.2 Some Properties of the Cell-Vertex Stencil . . . 105
- 7.3 Linearization of the Equations 106
- 7.4 Fourier Analysis of the Linearized Equations 107
- 7.5 Numerical Verification 111

7.6 Conclusions 113

8 Scramjet Inlets   120

    8.1 Introduction 120

    8.2 Two-Dimensional Test Cases 121

        8.2.1 $M_\infty = 5$, 0° Yaw 122

        8.2.2 $M_\infty = 5$, 5° Yaw 123

        8.2.3 $M_\infty = 2$, 0° Yaw 123

        8.2.4 $M_\infty = 3$, 0° Yaw 124

        8.2.5 $M_\infty = 3$, 7° Yaw 125

        8.2.6   Inlet Performance and Total Pressure Loss 125

    8.3 Three-Dimensional Results 126

9 Summary and Conclusions   138

    9.1 Summary 138

    9.2 Contributions of the Thesis 139

    9.3 Conclusions 141

    9.4 Areas for Further Exploration 142

A Computational Issues   144

    A.1 Introduction 144

    A.2 Vectorization and Parallelization Issues 144

    A.3 Computer Memory Requirements 146

        A.3.1   Two-Dimensional Memory Requirements 147

      A.3.2 Three-Dimensional Memory Requirements 147

  A.4 Data Structures for Adaptation 148

      A.4.1 Finding the Children of an Element 149

      A.4.2 Finding The Adjacent Element 150

B Scramjet Geometry Definition   152

References   154

List of Symbols   162

Index   164

Chapter 1
Introduction

Computational methods are playing an ever-increasing role in the design of flight vehicles [25, 62]. With the current computational power available, flow over (or in) realistic geometries can be calculated, if a computational grid can be generated. This need for geometric flexibility has given rise to unstructured grid algorithms, called finite element algorithms in this thesis. This thesis develops a new, adaptive finite element algorithm for the Euler equations describing the flow of an inviscid, compressible, ideal gas.

1.1 Research Goals

The research presented here has four main goals. The first goal is the development of an adaptive finite element solution algorithm for the solution of the steady Euler equations in two and three dimensions using quadrilateral and hexahedral elements. To the author's knowledge, this thesis presents the first use of adaptation by grid embedding using hexahedral elements in three dimensions. Hexahedral elements offer large CPU and memory reductions over tetrahedral elements, and quadrilateral elements have a similar, although less marked, advantage in two dimensions. Significant computational cost reductions for a given level of accuracy are obtained with the adaptive algorithm. The new features here are the use of a multistage time integration scheme and the development of the adaptive hexahedral element in three dimensions.

The second goal is the development of a biquadratic finite element for the two-dimensional Euler equations. A single biquadratic element should be more accurate than the four bilinear elements it would replace on a grid with a comparable number of nodes. While quadratic elements have been tried before with only limited success [9], this thesis develops and demonstrates the utility of the higher-order elements, and presents results indicating increased accuracy at reduced computational cost.

The third goal is to compare several unstructured grid numerical methods, the Galerkin method, the cell-vertex method, and the central difference method, and to show the differences between them. This will (hopefully) end some of the current confusion in the literature by demonstrating that many of the currently used algorithms are finite element methods in disguise, and allow discussion to be focused on the issues of structured grid vs. unstructured grid and algorithm selection.

The fourth goal is the explanation of the low wave number numerical errors which occur near regions of high gradient in many problems. A dispersion analysis is presented which allows one to predict both the frequency of the oscillations and their location as a function of physical and computational parameters such as Mach number, grid aspect ratio and solution algorithm.

1.2 Overview of Thesis

The thesis begins with a brief survey of current research in finite element methods as applied to the Euler equations. Chapter 2 introduces the governing equations and the boundary conditions used in this thesis. Chapter 3 presents some fundamental concepts of finite element methods, and derives the bilinear, biquadratic, and trilinear elements. Chapter 4 presents a complete description of the solution algorithm, and demonstrates that many other numerical methods can be cast into a finite element form. Chapter 5 presents some examples to verify the implementation of the numerical solver, and presents a comparison of three particular computational methods often used to solve the Euler equations. Two of these methods are shown to be robust, efficient methods for the solution of the Euler equations. Chapter 6 introduces the idea of adaptation, and presents examples showing how it can reduce the computational cost for a given level of accuracy, and how adaptation can reduce the sensitivity of a solution to a poor initial grid. Chapter 7 presents a dispersion analysis of the numerical methods, and introduces the concept of spatial group velocity. The spatial group velocity can be used to predict the location of certain types of low wave-number solution errors. Chapter 8 returns to the physical world and examines some of the interesting features of flows in scramjet inlets. Chapter 9 draws some conclusions from this research and suggests areas for future exploration. The thesis is completed with an appendix illustrating some of the computational issues in finite element methods, followed by appendices containing the listings of the computer codes.

1.3 Survey of Finite Element Methods for the Euler Equations

This section provides a brief review of the history of the finite element algorithm described in this thesis, and indicates other research going on in the field. The survey paper by Jameson [24] provides a good background for the history of numerical methods for the Euler equations. In 1954, Lax published a paper on the solution of hyperbolic equations with weak solutions [35]. Several years were spent laying the mathematical foundations for the solution procedures, and in the late 1960's and early 1970's the first Euler solvers began to emerge [46]. These early solvers were severely limited by the computational power available in those days. As a result, much of the CFD effort was based on the solution of the transonic full potential equations. As the 70's turned into the 80's, computer capabilities increased enough to make the solution of the Euler equations practical. The finite volume methods of Jameson [28] and Ni [51] date from this time. About this time, it also became possible to compute the flow over realistic two-dimensional geometries, and unstructured grid ideas began to migrate from structural mechanics into fluid mechanics. Much of the pioneering work on unstructured grids was done by French researchers at INRIA and Dassault [7]. In the mid 1980's, several unstructured algorithms began to emerge, and as newer computer architectures with larger memories and hardware scatter/gather became available, more researchers shifted into the unstructured mesh arena, some with "finite volume" methods, and others with "finite element" methods. At present, finite element methods are gaining widespread acceptance in fluid mechanics, and many researchers are investigating the use of unstructured grids for the solution of realistic problems.

There are many different formulations of the finite element method in the literature. Finite element algorithms different from the algorithm used in this thesis include the Petrov-Galerkin method (in which the test functions change with the solution to produce an oscillation-free solution) [18], the Euler-Taylor-Galerkin method (essentially a Lax-Wendroff type method with a finite element spatial discretization) [6, 10], Clebsch-transformed variables (corrections to the potential equations) [12] and various direct methods (Newton solvers) [8, 15].

A sampling of other researchers in finite element methods includes Hughes, et al. at Stanford [19, 17], Kennon [30] and Oden [52] at the University of Texas, Löhner [43, 40, 39] at the Naval Research Lab, a group in Virginia at NASA Langley and Old Dominion University [6, 58, 71], Jameson from Princeton [27], a group at Swansea [48, 53, 54], groups at INRIA and Dassault [3, 55, 67], and several groups in Japan [50, 66]. Although the term finite element is not used, the work of Dannenhoffer [20, 22], can also be considered a finite element method. Many other groups throughout the world are also turning their attention to the development of unstructured grid methods for the Euler and Navier-Stokes equations [41]. The current interest in finite element and unstructured grid computational methods shows no signs of abating, and more researchers enter the field every year.

Chapter 2
Governing Equations

This chapter introduces the governing equations used throughout the remainder of the thesis. The non-dimensionalization of the equations is discussed, and the boundary conditions are described. Finally, some of the limitations of the equations are discussed.

2.1 Euler Equations

The equations solved are the Euler equations describing the flow of a compressible, inviscid, ideal gas. For these equations to hold, the following assumptions are necessary:

- The fluid is a homogeneous continuum;
- The Reynolds number is infinite (inviscid);
- The Peclet number is infinite (non-conducting);
- The fluid obeys the ideal gas law.

While these assumptions do not hold exactly for any real flow, for a large class of problems they are a very good approximation. This thesis considers solutions to the steady-state Euler equations, but since the solution method involves a pseudo-time marching scheme the unsteady Euler equations are described here. Also, no body forces are considered in these equations.

The Euler equations in three dimensions can be written in conservation form as

$$\frac{\partial}{\partial t}\begin{bmatrix}\rho\\\rho u\\\rho v\\\rho w\\\rho e\end{bmatrix}+\frac{\partial}{\partial x}\begin{bmatrix}\rho u\\\rho u^2+p\\\rho uv\\\rho uw\\\rho uh_0\end{bmatrix}+\frac{\partial}{\partial y}\begin{bmatrix}\rho v\\\rho uv\\\rho v^2+p\\\rho vw\\\rho vh_0\end{bmatrix}+\frac{\partial}{\partial z}\begin{bmatrix}\rho w\\\rho uw\\\rho vw\\\rho w^2+p\\\rho wh_0\end{bmatrix}=0, \quad (2.1)$$

where where e is total energy, p is pressure, ρ is density, u, v, and w are the flow velocities in the x, y, and z directions, and h_0 is the total enthalpy, given by the thermodynamic relation

$$h_0 = e + \frac{p}{\rho}. \quad (2.2)$$

In addition, one requires an equation of state in order to complete the set of equations. For an ideal gas, this can be written

$$\frac{p}{\rho} = (\gamma - 1)\left[e - \frac{1}{2}(u^2 + v^2 + w^2)\right], \quad (2.3)$$

where the specific heat ratio $\gamma = 1.4$ is constant for all calculations reported.

It is convenient to write the equations in vector form as

$$\frac{\partial \mathbf{U}}{\partial t} + \frac{\partial \mathbf{F}}{\partial x} + \frac{\partial \mathbf{G}}{\partial y} + \frac{\partial \mathbf{H}}{\partial z} = 0, \quad (2.4)$$

where $\mathbf{U}$ is the vector of state variables and $\mathbf{F}$, $\mathbf{G}$, and $\mathbf{H}$ are flux vectors in the x, y, and z directions, corresponding to the vectors in Eq. (2.1) above. To restrict these equations to two dimensions, drop the z derivatives and the z momentum equation.

The Euler equations described above provide a complete description of the compressible flow of an inviscid, non-conducting, ideal gas in the absence of body forces. For many problems this is a reasonable set of restrictions, but there are flows of interest where the Euler equations will not suffice. Each of the assumptions involved in the formation of the Euler equations will be examined.

The assumption that the fluid is continuous and homogeneous can break down for very low density flows, such as flow in the upper atmosphere or flows involving the mixing of multiple components. The

Table 2.1: Scaling Factors for Non-Dimensionalization

Variable	Factor	Free Stream Value
u, v, w	a_∞	$M_{X\infty}, M_{Y\infty}, M_{Z\infty}$
ρ	ρ_∞	1
p	$\rho_\infty a_\infty^2$	$1/\gamma$
e, h	a_∞^2	$M_\infty^2/2 + 1/\gamma(\gamma-1), M_\infty^2/2 + 1/(\gamma-1)$
x, y, z	L	--
t	L/a_∞	--

assumption that the fluid is inviscid means that the Euler equations are inadequate if one is interested in viscous phenomena such as skin friction, boundary layers, separation and stall, or viscous-inviscid interaction. The non-conducting assumption means that heat transfer problems cannot be modeled. In many cases the ideal gas law breaks down. These can include problems in hypersonics in which the gas can dissociate and/or excite additional internal energy modes, combustion and other chemically reacting flows, and free-molecule flow (this also violates the first assumption). Finally, flows in which body forces are important (weather prediction or magnetohydrodynamics, for example) require the addition of additional terms to the equations.

2.2 Non-Dimensionalization of the Equations

It is often convenient to non-dimensionalize the governing equations for a problem, since this clarifies the scales important to a problem, makes solutions independent of any particular system of units, and often helps reduce the sensitivity of a numerical solution to round-off errors. Table 2.2 lists the scaling factors for each of the problem variables. With this non-dimensionalization, the Euler equations become

$$\frac{\partial}{\partial t'}\begin{bmatrix}\rho' \\ \rho'u' \\ \rho'v' \\ \rho'w' \\ \rho'e'\end{bmatrix} + \frac{\partial}{\partial x'}\begin{bmatrix}\rho'u' \\ \rho'u'^2+p' \\ \rho'u'v' \\ \rho'u'w' \\ \rho'u'h_0'\end{bmatrix} + \frac{\partial}{\partial y'}\begin{bmatrix}\rho'v' \\ \rho'u'v' \\ \rho'v'^2+p' \\ \rho'v'w' \\ \rho'v'h_0'\end{bmatrix} + \frac{\partial}{\partial z'}\begin{bmatrix}\rho'w' \\ \rho'u'w' \\ \rho'v'w' \\ \rho'w'^2+p' \\ \rho'w'h_0'\end{bmatrix} = 0, \quad (2.5)$$

where the ′ variables are non-dimensional. This non-dimensionalization shows that the Euler equations have two associated non-dimensional parameters, which enter through the boundary conditions and the state equation: the Mach number M and the ratio of specific heats γ. Note that the non-dimensional parameters associated with the Euler equations do not appear in Eq. (2.5). Since this is the case, all further discussions will be based on the non-dimensional variables, and the ′ superscript will be dropped.

2.3 Auxiliary Quantities

It is convenient to define a number of auxiliary physical quantities in terms of the primitive quantities $\rho, u, v, w,$ and p. These are the following:

Local speed of sound: $\quad a = \sqrt{\dfrac{\gamma p}{\rho}}$,

Mach Number: $\quad M = \dfrac{\sqrt{u^2 + v^2 + w^2}}{a}$,

Total Pressure: $\quad p_0 = p(1 + \dfrac{\gamma - 1}{2} M^2)^{\gamma/(\gamma-1)}$,

Total Pressure Loss: $\quad p_{\text{loss}} = \dfrac{p_{0\infty} - p_0}{p_{0\infty}}$,

Entropy: $\quad \Delta S = \log \dfrac{\gamma p}{\rho^\gamma}$,

where the free stream entropy is defined to be 0.

2.4 Boundary Conditions

In order to solve any set of differential equations, boundary conditions need to be specified. In this thesis, two types of boundaries are defined for the Euler equations: solid surfaces and "open" boundaries. The implementation of these boundary conditions is discussed in Section 4.4.

2.4.1 Solid Surface Boundary Conditions

At a solid surface boundary, there is no mass flux normal to the boundary. This is equivalent to saying

$$\vec{u} \cdot \hat{n} = 0, \qquad (2.6)$$

where $\vec{u}$ is the velocity vector and $\hat{n}$ is the unit normal to the surface.

2.4.2 Open Boundary Conditions

The "open" or far-field boundary conditions are based on quasi-one-dimensional characteristic theory. The three-dimensional Euler equations are transformed into a system based on coordinates normal to the boundary, and derivatives tangential to the boundaries are neglected. The resulting equations are diagonalized assuming locally isentropic flow, yielding the characteristic equations

$$\frac{\partial Q}{\partial t} + (u_\xi + a)\frac{\partial Q}{\partial \xi} = 0, \qquad (2.7)$$

$$\frac{\partial R}{\partial t} + (u_\xi - a)\frac{\partial R}{\partial \xi} = 0, \qquad (2.8)$$

$$\frac{\partial u_\eta}{\partial t} + u_\xi \frac{\partial u_\eta}{\partial \xi} = 0, \qquad (2.9)$$

$$\frac{\partial u_\zeta}{\partial t} + u_\xi \frac{\partial u_\zeta}{\partial \xi} = 0, \qquad (2.10)$$

$$\frac{\partial S}{\partial t} + u_\xi \frac{\partial S}{\partial \xi} = 0, \qquad (2.11)$$

where (ξ, η, ζ) are the transformed directions, with ξ normal to the boundary, S is the entropy, u_η and u_ζ are the velocities tangential to the boundary, and Q and R are the Riemann invariants [36]

$$Q = u_\xi + \frac{2a}{\gamma - 1}, \qquad (2.12)$$

$$R = u_\xi - \frac{2a}{\gamma - 1}, \qquad (2.13)$$

derived from the diagonalized system. If there is no entropy variation normal to the boundary, these invariants are exact, otherwise they are approximate. These equations are decoupled wave equations, and so these characteristic variables are convected normal to the boundary in a direction determined by the sign of the associated wave velocity. For example, if $0 < u_\xi < a$, the boundary is a subsonic inflow boundary, so Q, S, u_η and u_ζ propagate into the domain, while R propagates out of the domain.

Chapter 3
Finite Element Fundamentals

This chapter introduces some of the important concepts in finite element methods. The terms *element, node, edge* and *face* are defined, and the transformations between physical and computational space are described. A discussion of derivative calculation is given, and the 4-node, 2-D bilinear, the 9-node, 2-D biquadratic, and the 8-node, 3-D trilinear elements are developed.

3.1 Basic Definitions

The finite element method subdivides the physical domain of interest into small subdomains called *elements,* each of which is composed of some number of *nodes.* Figure 3.1 shows how a domain might be divided into elements, in this case six. The nodes are indicated by the black circles. Note that not all the elements are made up of the same numbers of nodes. For example, the mesh shown has pentagonal, quadrilateral and triangular elements. The finite element method does not restrict one to identical elements, in general. Quadrilateral and triangular elements in two dimensions have been combined in a single problem, by Ramakrishnan, for example [58]. In actual practice one usually only uses a few different types of elements in a particular problem. For a real problem, a domain will typically be divided into hundreds, thousands, or hundreds of thousands of elements.

Elements also have *faces*, defined to be the structures of dimension one less than the element dimension and composed of element intersections. An *edge* is the intersection of faces. Note that in two dimensions, an edge and a node are the same thing, but in three dimensions they are not. Figure 3.2 shows faces and edges in two and three dimensions.

In this thesis, quadrilateral elements are used in two dimensions and hexahedral elements are used in three dimensions. The algorithm it-

6 Elements, 10 Nodes

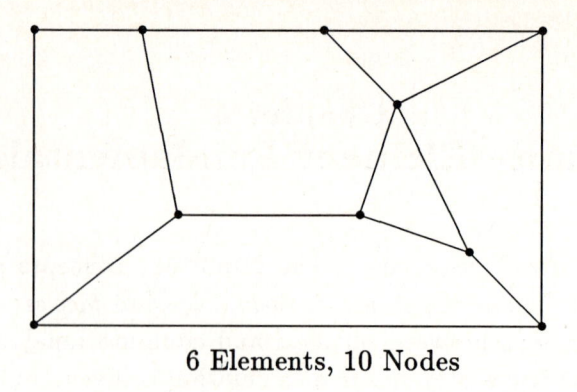

Figure 3.1: Example of a General Finite Element Discretization

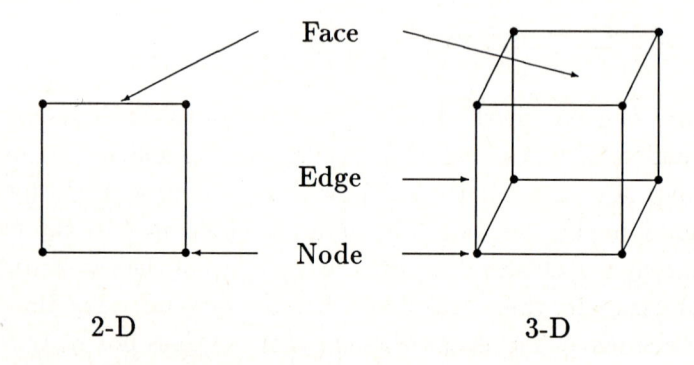

2-D   3-D

Figure 3.2: Definition of Finite Element Terms

self is applicable to arbitrary polygons and polyhedra. The advantage of quadrilateral and hexahedral elements is that for a given number of nodes, a triangular mesh will have roughly twice as many elements as a quadrilateral mesh, and in three dimensions, a tetrahedral mesh will have roughly five times as many elements as a hexahedral mesh. Since there are many operations that are performed on elements, there is a significant potential for memory and CPU savings with reduced numbers of elements. The disadvantages of the quadrilateral and hexahedral elements are that grid generation may be more difficult for some problems, and where grid embedding is used, there is the problem of interface treatment. On the other hand, the formulation of the finite element method permits the use of degenerate elements, that is, elements in which one

Node Numbering: 1-2-3-1

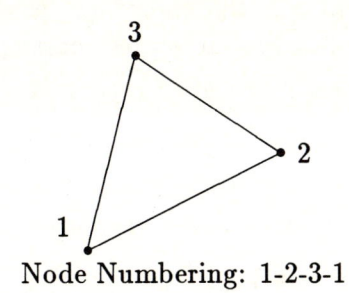

Figure 3.3: Quadrilateral Element Degenerated into a Triangular Element

or more nodes are repeated to form an element with a smaller effective number of nodes. Figure 3.3 shows how a quadrilateral element can be degenerated into a triangular element by repeating node 1, for example. At this point it is important to emphasize the *unstructured* nature of the finite element method. In the finite element method, all operations are done at the element level, with element contributions assembled (distributed) to the nodes. Computationally, there are no structures such as grid lines, although one may see something that looks like a grid line in a picture of a mesh. This distinction is what give finite element methods their flexibility. It is not necessary for each grid point to be indexed by (i,j,k) or by some similar scheme, and a node may belong to any number of elements.

A vast literature exists describing finite element, finite volume and finite difference algorithms for various equations. As will be shown in Section 4.3, many of the finite volume and finite difference algorithms can be viewed as finite element algorithms, so the real distinction should not be one of *name* (finite element, volume or difference) but of substance (structured mesh or unstructured mesh, what the difference stencil actually looks like, etc.)

3.2 Finite Elements and Natural Coordinates

The finite element method provides a way to make a convenient transformation between a local, computational space (natural coordinates) and a global, physical space. The following discussion will emphasize this in two dimensions, but the ideas are identical in three dimensions (or even in one dimension).

In the finite element discretization one assumes that within each element, some quantity $q^{(e)}$ is determined by its nodal values q_i and a set of *shape* or *interpolation functions* $N_i^{(e)}$ so that

$$q^{(e)} = \sum_{i=1}^{m} N_i^{(e)} q_i, \qquad (3.1)$$

where m is the number of nodes in the element. The elemental shape functions are summed to give global shape functions $\mathbf{N}_i$, so that globally q can be written

$$q(x,y) = \sum_{i=1}^{M} \mathbf{N}_i(x,y) q_i, \qquad (3.2)$$

where M is the total number of nodes in the mesh.

3.2.1 Properties of Interpolation Functions

The interpolation functions $N_i^{(e)}$ (also called shape functions or *trial functions*) must have certain properties in order for the finite element approximation to be valid. These properties are as follows:

1. The shape function $N_i^{(e)}$ must be 1 at node i and 0 at all other nodes of the domain. This is required so that the relation in Eq. (3.2) can hold for each node.

2. The shape function $N_i^{(e)}$ should be 0 outside of element e. Strictly speaking, this is only required if one desires a local finite element approximation. All the shape functions used in this report are local and satisfy this property. A consequence of this property is that the global shape function $\mathbf{N}_i$ at node i is just a union of the elemental shape functions $N_i^{(e)}$ for all the elements containing node i.

3. In each element, the sum of all the nodal shape functions $N_i^{(e)}$ should be identically 1. This is so the constant function can be represented exactly (a requirement for consistency in the approximation).

There are two things to note about these requirements. First, requirements 2 and 3 imply that constant functions can be represented exactly for all geometries. Second, these requirements do not force the shape functions to be continuous between the elements, and some researchers have made use of discontinuous trial functions in their formulations [2, 30]. In this report, all interpolation functions will be continuous in the elements and across the element boundaries.

Typically, interpolation functions are chosen to be polynomials in some natural coordinates (ξ, η). The degree of the polynomial approximation is related to the accuracy of the interpolation desired and to the specifics of the problem. For example, this author has found that for many structural mechanics problems, biquadratic interpolation functions give better results than bilinear interpolation functions [65], but in the solution of the Euler equations the question of the optimal order of shape function is still an open one.

3.2.2 Natural Coordinates and Derivative Calculation

Interpolation functions are usually chosen to be polynomials in some natural coordinate system (ξ, η). Strang [69] indicates that polynomials are the optimal choice for interpolation functions in the sense that in order to obtain an kth order accurate approximation to an sth derivative on a regular mesh, the interpolation functions must be complete (be able to represent exactly) all polynomials of degree $k + s - 1$. Thus, a set of polynomials will have the smallest number of elements for a given order of accuracy.

The geometry of the element is interpolated in terms of nodal coordinates. That is, one states that within an element,

$$x^{(e)} = \sum N_i^{(e)}(\xi, \eta) x_i, \qquad (3.3)$$
$$y^{(e)} = \sum N_i^{(e)}(\xi, \eta) y_i, \qquad (3.4)$$

where x_i and y_i are the coordinates of node i in element e. For simplicity, all derivations are shown in two dimensions, and the extension to three dimensions is straightforward.

It is useful to be able to write the derivative of a quantity in terms of the nodal values of that quantity. Since the derivatives are usually desired in physical space, we require the Jacobian of the transformation. One can write:

$$\begin{bmatrix} \dfrac{\partial}{\partial \xi} \\ \dfrac{\partial}{\partial \eta} \end{bmatrix} = J \begin{bmatrix} \dfrac{\partial}{\partial x} \\ \dfrac{\partial}{\partial y} \end{bmatrix}, \qquad (3.5)$$

where J is the Jacobian matrix (valid within each element)

$$J = \begin{bmatrix} \dfrac{\partial x}{\partial \xi} & \dfrac{\partial y}{\partial \xi} \\ \dfrac{\partial x}{\partial \eta} & \dfrac{\partial y}{\partial \eta} \end{bmatrix} = \begin{bmatrix} \sum x_i \dfrac{\partial N_i^{(e)}}{\partial \xi} & \sum y_i \dfrac{\partial N_i^{(e)}}{\partial \xi} \\ \sum x_i \dfrac{\partial N_i^{(e)}}{\partial \eta} & \sum y_i \dfrac{\partial N_i^{(e)}}{\partial \eta} \end{bmatrix}. \qquad (3.6)$$

When J is known (and non-singular), J^{-1} can be calculated, so one can write the derivatives of a quantity q in each element as follows:

$$\begin{bmatrix} \dfrac{\partial q}{\partial x} \\ \dfrac{\partial q}{\partial y} \end{bmatrix} = J^{-1} \begin{bmatrix} \sum_i q_i \dfrac{\partial N_i^{(e)}}{\partial \xi} \\ \sum_i q_i \dfrac{\partial N_i^{(e)}}{\partial \eta} \end{bmatrix}, \qquad (3.7)$$

where the q_i are the nodal values of q. Note that this requires the Jacobian to be non-singular for all ξ and η in the element. For the bilinear transformation, this will be true if, and only if, the element is convex in physical coordinates. To see this, note that $|J|$ is linear in an element, so if the sign of $|J|$ changes between two nodes, $|J|$ will be zero somewhere in the interior. If the element is non-convex, $|J|$ will have a different sign at the node where the interior angle exceeds 180°. The complete proof is given by Strang [69].

If the same shape functions are used to interpolate both the element geometry and the quantity q, the element is called an *isoparametric* element. If the shape functions used to interpolate the geometry are of a lesser degree than the interpolation functions for the quantity q, the element is termed *subparametric*. In this thesis, isoparametric bi- and trilinear elements and subparametric biquadratic elements are used.

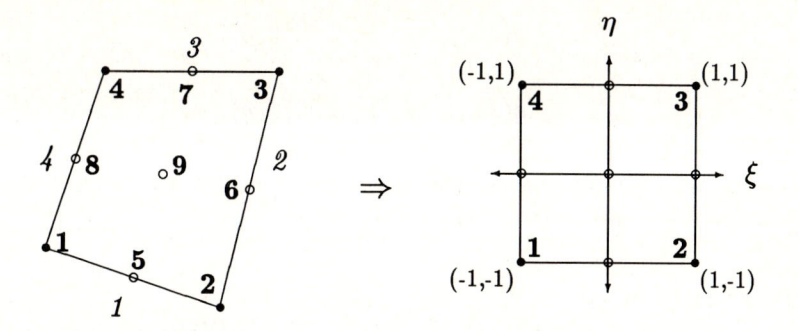

Physical Coordinates   Natural Coordinates

Node numbering in **bold**, face numbering in *italic*

Figure 3.4: Geometry of Two-Dimensional Element

3.3 Typical Elements

Figure 3.4 shows the geometry for the 4-node bilinear and 9-node biquadratic elements in both physical and natural coordinates. This figure also shows the node and face numberings used at the element level. The open circles indicate nodes that are present only in the 9-node element. Figure 3.5 (on page 20) shows the equivalent information for the three-dimensional, trilinear element. To help clarify this figure, Table 3.1 lists the nodes making up each face of the element.

3.3.1 Bilinear Element

The section presents the shape functions for the bilinear, 4-node element and gives the explicit formulas for the Jacobian and its inverse. See Fig. 3.4 for the geometry of the element. In natural coordinates, the nodal interpolation functions are

$$N_1 = (1-\xi)(1-\eta)/4, \quad (3.8a)$$
$$N_2 = (1+\xi)(1-\eta)/4, \quad (3.8b)$$
$$N_3 = (1+\xi)(1+\eta)/4, \quad (3.8c)$$
$$N_4 = (1-\xi)(1+\eta)/4, \quad (3.8d)$$

and one can write

$$x(\xi,\eta) = \sum_{i=1}^{4} x_i N_i(\xi,\eta), \quad (3.9a)$$

$$y(\xi,\eta) = \sum_{i=1}^{4} y_i N_i(\xi,\eta), \qquad (3.9b)$$

$$q(\xi,\eta) = \sum_{i=1}^{4} q_i N_i(\xi,\eta), \qquad (3.9c)$$

where x_i and y_i are the coordinates of the nodes and q_i are the nodal values of some quantity q. It will be convenient to expand these quantities:

$$\begin{aligned}
x(\xi,\eta) &= a_1 + a_2\xi + a_3\eta + a_5\xi\eta, & (3.10a) \\
y(\xi,\eta) &= b_1 + b_2\xi + b_3\eta + b_5\xi\eta, & (3.10b) \\
q(\xi,\eta) &= c_1 + c_2\xi + c_3\eta + c_5\xi\eta, & (3.10c)
\end{aligned}$$

where the coefficients a_1, a_2, a_3, a_5 are

$$\begin{aligned}
a_1 &= (x_1 + x_2 + x_3 + x_4)/4, & (3.11a) \\
a_2 &= (-x_1 + x_2 + x_3 - x_4)/4, & (3.11b) \\
a_3 &= (-x_1 - x_2 + x_3 + x_4)/4, & (3.11c) \\
a_5 &= (x_1 - x_2 + x_3 - x_4)/4, & (3.11d)
\end{aligned}$$

and similarly for b and c. The $\xi\eta$ term has the subscript "5" instead of the subscript "4" to reserve the subscript "4" for the ξ^2 term in the biquadratic expansions following.

Now the Jacobian can be formed. Writing in terms of the a's and b's,

$$J = \begin{bmatrix} a_2 + a_5\eta & b_2 + b_5\eta \\ a_3 + a_5\xi & b_3 + b_5\xi \end{bmatrix}, \qquad (3.12)$$

$$|J| = (a_2 b_3 - a_3 b_2) + (a_2 b_5 - a_5 b_2)\xi + (a_5 b_3 - a_3 b_5)\eta, \qquad (3.13)$$

$$J^{-1} = \frac{1}{|J|} \begin{bmatrix} b_3 + b_5\xi & -(b_2 + b_5\eta) \\ -(a_3 + a_5)\xi & a_2 + a_5\eta \end{bmatrix}, \qquad (3.14)$$

so derivatives (and integrals) of quantities can now be calculated in physical coordinates.

3.3.2 Biquadratic Element

The biquadratic element used is a subparametric, 9-node element. The geometry is interpolated exactly as the bilinear element just described. This section presents the analogs to Eqs. (3.8) and (3.11). The biquadratic shape functions are

$$N_1 = \xi(1-\xi)\eta(1-\eta)/4, \tag{3.15a}$$
$$N_2 = -\xi(1+\xi)\eta(1-\eta)/4, \tag{3.15b}$$
$$N_3 = \xi(1+\xi)\eta(1+\eta)/4, \tag{3.15c}$$
$$N_4 = -\xi(1-\xi)\eta(1+\eta)/4, \tag{3.15d}$$
$$N_5 = -(1-\xi^2)\eta(1-\eta)/2, \tag{3.15e}$$
$$N_6 = \xi(1-\xi)(1+\eta^2)/2, \tag{3.15f}$$
$$N_7 = (1-\xi^2)\eta(1+\eta)/2, \tag{3.15g}$$
$$N_8 = -\xi(1-\xi)(1+\eta^2)/2, \tag{3.15h}$$
$$N_9 = (1-\xi^2)(1-\eta^2), \tag{3.15i}$$

so that one can write some quantity q as

$$q = c_1 + c_2\xi + c_3\eta + c_4\xi^2 + c_5\xi\eta + c_6\eta^2 + c_7\xi^2\eta + c_8\xi\eta^2 + c_9\xi^2\eta^2, \tag{3.16}$$

where

$$c_1 = q_9, \tag{3.17a}$$
$$c_2 = (q_6 - q_8)/2, \tag{3.17b}$$
$$c_3 = (q_7 - q_5)/2, \tag{3.17c}$$
$$c_4 = (q_6 + q_8 - 2q_9)/2, \tag{3.17d}$$
$$c_5 = (q_1 - q_2 + q_3 - q_4)/4, \tag{3.17e}$$
$$c_6 = (q_5 + q_7 - 2q_9)/2, \tag{3.17f}$$
$$c_7 = (-q_1 - q_2 + q_3 + q_4 + 2q_5 - 2q_7)/4, \tag{3.17g}$$
$$c_8 = (-q_1 + q_2 + q_3 - q_4 - 2q_6 - 2q_8)/4, \tag{3.17h}$$
$$c_9 = (q_1 + q_2 + q_3 + q_4 - 2q_5 - 2q_6 - 2q_7 - 2q_8 + 4q_9)/4, \tag{3.17i}$$

and the q_i are the nodal values of the quantity q. Since the element is subparametric, the Jacobians are identical to Eqs. (3.12), (3.13), and (3.14).

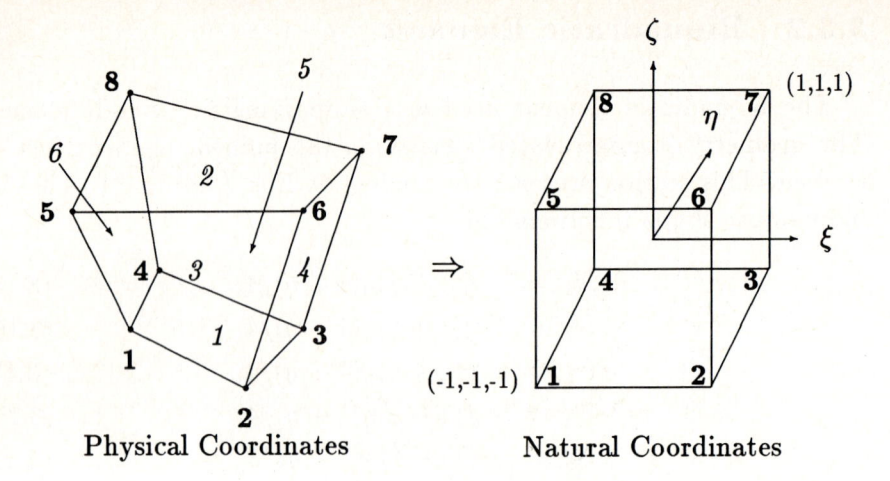

Figure 3.5: Geometry of Three-Dimensional Element

Physical Coordinates → Natural Coordinates

Nodes in **bold**, Faces in *italic*

Table 3.1: Nodes for Each Face, Trilinear Element

Face	Nodes on Face	Face	Nodes on Face
1	1-2-3-4	4	2-3-7-6
2	5-6-7-8	5	4-3-7-8
3	1-2-6-5	6	1-4-8-5

3.3.3 Trilinear Element

The 8-node, three-dimensional element shown in Fig. 3.5 is a trilinear element. To help clarify this figure, Table 3.1 lists the nodes making up each face of the element. The shape functions are

$$N_1 = (1-\xi)(1-\eta)(1-\zeta)/8, \quad (3.18a)$$
$$N_2 = (1+\xi)(1-\eta)(1-\zeta)/8, \quad (3.18b)$$
$$N_3 = (1+\xi)(1+\eta)(1-\zeta)/8, \quad (3.18c)$$
$$N_4 = (1-\xi)(1+\eta)(1-\zeta)/8, \quad (3.18d)$$
$$N_5 = (1-\xi)(1-\eta)(1+\zeta)/8, \quad (3.18e)$$
$$N_6 = (1+\xi)(1-\eta)(1+\zeta)/8, \quad (3.18f)$$
$$N_7 = (1+\xi)(1+\eta)(1+\zeta)/8, \quad (3.18g)$$

$$N_8 = (1-\xi)(1+\eta)(1+\zeta)/8, \tag{3.18h}$$

and one can write

$$q = d_1 + d_2\xi + d_3\eta + d_4\zeta + d_5\xi\eta + d_6\eta\zeta + d_7\xi\zeta + d_8\xi\eta\zeta, \tag{3.19}$$

where

$$d_1 = (q_1 + q_2 + q_3 + q_4 + q_5 + q_6 + q_7 + q_8)/8, \tag{3.20a}$$
$$d_2 = (-q_1 + q_2 + q_3 - q_4 - q_5 + q_6 + q_7 - q_8)/8, \tag{3.20b}$$
$$d_3 = (-q_1 - q_2 + q_3 + q_4 - q_5 - q_6 + q_7 + q_8)/8, \tag{3.20c}$$
$$d_4 = (-q_1 - q_2 - q_3 - q_4 + q_5 + q_6 + q_7 - q_8)/8, \tag{3.20d}$$
$$d_5 = (q_1 - q_2 + q_3 - q_4 + q_5 - q_6 + q_7 + q_8)/8, \tag{3.20e}$$
$$d_6 = (q_1 + q_2 - q_3 - q_4 - q_5 - q_6 + q_7 + q_8)/8, \tag{3.20f}$$
$$d_7 = (q_1 - q_2 - q_3 + q_4 - q_5 + q_6 + q_7 - q_8)/8, \tag{3.20g}$$
$$d_8 = (-q_1 + q_2 - q_3 + q_4 + q_5 - q_6 + q_7 - q_8)/8. \tag{3.20h}$$

The Jacobians are calculated in a similar manner as those above. Due to the complexity of the expressions involved, only the Jacobian matrix itself is shown. The three-dimensional Jacobian J is

$$J = \begin{bmatrix} a_2 + a_5\eta + a_7\zeta + a_8\eta\zeta & b_2 + b_5\eta + b_7\zeta + b_8\eta\zeta & c_2 + c_5\eta + c_7\zeta + c_8\eta\zeta \\ a_3 + a_5\xi + a_6\zeta + a_8\xi\zeta & b_3 + b_5\xi + b_6\zeta + b_8\xi\zeta & c_3 + c_5\xi + c_6\zeta + c_8\xi\zeta \\ a_4 + a_6\eta + a_7\xi + a_8\xi\eta & b_4 + b_6\eta + b_7\xi + b_8\xi\eta & c_4 + c_6\eta + c_7\xi + c_8\xi\eta \end{bmatrix}, \tag{3.21}$$

where a_i, b_i, and c_i are the coefficients in the expansions of x, y, and z in the element.

Chapter 4
Solution Algorithm

This chapter describes in detail the finite element solution algorithm for the Euler equations. The application of the finite element method to the spatial discretization is described, and section 4.3 introduces the Galerkin finite element, "cell-vertex" finite element and "central difference" finite element methods. The implementation of boundary conditions is discussed in section 4.4. All of these methods require added damping for stability, and this is discussed in section 4.5. Section 4.6 describes the pseudo-time marching method. Finally, section 4.7 describes the conditions on the test and trial functions needed to obtain consistency and conservation.

4.1 Overview of Algorithm

This section describes briefly the steps taken in the solution of a problem; each step is discussed in detail in the following sections. The steady-state Euler equations are solved using a pseudo-time marching technique. This means that from some initial condition, the solution is evolved by an iterative technique resembling the solution of the unsteady problem until it stops changing. This time marching consists of three steps. First, a residual representative of the difference between the steady solution and the current solution is calculated. Second some additional damping terms are added to this residual. Third, the current solution is updated to obtain the next approximation. This process is repeated until the desired degree of convergence is obtained. Convergence is signaled when the RMS of all changes divided by the RMS of all the state vectors is less than some specified number, usually around 10^{-5}. Other norms are possible, but the differences in the solutions produced by different indicators are not significant.

A new contribution is the use of the four-step multistage time integration scheme. Previous work has often used a two-step Lax-Wendroff time integration method [6, 43], but Ramakrishnan, Bey and Thornton have shown that the multistage time integration method developed herein has better stability properties than the two-step Lax-Wendroff time integration method when used with adaptive meshes [58].

4.2 Spatial Discretization

The spatial discretization method begins with the Euler equations in conservation law form (Eq. (2.4)) written

$$\frac{\partial \mathbf{U}}{\partial t} + \frac{\partial \mathbf{F}}{\partial x} + \frac{\partial \mathbf{G}}{\partial y} + \frac{\partial \mathbf{H}}{\partial z} = 0, \qquad (4.1)$$

where $\mathbf{U}$ is the vector of state variables and $\mathbf{F}$, $\mathbf{G}$, and $\mathbf{H}$ are flux vectors in the x, y, and z directions. Within each element the state vector $\mathbf{U}^{(e)}$ and flux vectors $\mathbf{F}^{(e)}$, $\mathbf{G}^{(e)}$ and $\mathbf{H}^{(e)}$ are written

$$\mathbf{U}^{(e)} = \sum N_i^{(e)} \mathbf{U}_i, \qquad (4.2)$$

$$\mathbf{F}^{(e)} = \sum N_i^{(e)} \mathbf{F}_i, \qquad (4.3)$$

$$\mathbf{G}^{(e)} = \sum N_i^{(e)} \mathbf{G}_i, \qquad (4.4)$$

$$\mathbf{H}^{(e)} = \sum N_i^{(e)} \mathbf{H}_i, \qquad (4.5)$$

where $\mathbf{U}_i, \mathbf{F}_i, \mathbf{G}_i$ and $\mathbf{H}_i$ are the nodal values of the state vector and flux vectors, and $N_i^{(e)}$ is the set of interpolation functions for element e.

These expressions can be differentiated to obtain a formula for the derivative in each element in terms of the nodal values as described in Section 3.2.2. In all the following steps, the two-dimensional algorithm will be shown for simplicity. The steps are identical for three dimensions, with the $\mathbf{H}$ fluxes and z derivatives included.

The expression for the derivatives is substituted into equation (4.1) and summed over all elements to obtain

$$\begin{aligned}
\mathbf{N}_i \frac{d\mathbf{U}_i}{dt} &= -\frac{\partial \mathbf{N}_i}{\partial x}\mathbf{F}_i - \frac{\partial \mathbf{N}_i}{\partial y}\mathbf{G}_i \qquad (4.6) \\
&= -(J_{1,1}^{-1}\frac{\partial \mathbf{N}_i}{\partial \xi} + J_{1,2}^{-1}\frac{\partial \mathbf{N}_i}{\partial \eta})\mathbf{F}_i - (J_{2,1}^{-1}\frac{\partial \mathbf{N}_i}{\partial \xi} + J_{2,2}^{-1}\frac{\partial \mathbf{N}_i}{\partial \eta})\mathbf{G}_i
\end{aligned}$$

where $\mathbf{N}_i$ is now a global *row* vector of interpolation functions, determined by summing the interpolation functions for each element.

It is impossible to make Equation (4.6) hold for all points in space (since the space of interpolation functions does not include all solutions to the Euler equations), so some "average" solution is required. The next step creates a *weak* form of the equations. This can be thought of as a projection onto the space spanned by some other row vector of functions $\tilde{\mathbf{N}}$, called *test functions*, such that the error in the discretization is orthogonal to the space spanned by the test functions. In the weak form, the equation is no longer required to be satisfied pointwise, but instead the equation is required to hold for each test function. This allows the introduction of discontinuous solutions, as well as providing some means for obtaining the nodal values of the unknowns. For more detail on the mathematics involved see Strang's books [68, 69]. To create this weak form, premultiply Eq. (4.6) by $\tilde{\mathbf{N}}^T$ and integrate over the entire domain. When this is done, one obtains

$$M\frac{d\mathbf{U}_i}{dt} = -\iint (\tilde{\mathbf{N}}^T \frac{\partial \mathbf{N}}{\partial x}\mathbf{F}_i + \tilde{\mathbf{N}}^T \frac{\partial \mathbf{N}}{\partial y}\mathbf{G}_i)\, dV, \qquad (4.7)$$

$$M = \iint \tilde{\mathbf{N}}^T \mathbf{N}\, dV, \qquad (4.8)$$

which results in the semi-discrete equation

$$M\frac{d\mathbf{U}_i}{dt} = -R_x \mathbf{F}_i - R_y \mathbf{G}_i, \qquad (4.9)$$

where M is the consistent mass matrix, and R_x and R_y are residual matrices. The matrices M, R_x and R_y involve the integration of quantities over the domain. These integrations are done at the element level in natural coordinates, and assembled to give the global matrices. The selection of $\tilde{\mathbf{N}}$ is discussed in detail in section 4.3, but for now it is sufficient to note that each $\tilde{\mathbf{N}}$ is a polynomial in natural coordinates. In the calculation of the residual matrices, this results in the integration of a polynomial in (ξ, η) over the domain, because the Jacobian determinant in the denominator of Eq. (3.7) cancels the Jacobian determinant in the integration. To make this cancellation clear, consider the calculation of the R_x matrix:

$$\begin{aligned}
R_x^{(e)} &= \iint \tilde{\mathbf{N}}^T \frac{\partial \mathbf{N}}{\partial x} dx\, dy \qquad (4.10)\\
&= \iint \tilde{\mathbf{N}}^T (J_{1,1}^{-1} \frac{\partial \mathbf{N}}{\partial \xi} + J_{1,2}^{-1} \frac{\partial \mathbf{N}}{\partial \eta})|J| d\xi\, d\eta\\
&= \int_{-1}^{1}\int_{-1}^{1} \tilde{\mathbf{N}}^T (J_{1,1}^* \frac{\partial \mathbf{N}}{\partial \xi} + J_{1,2}^* \frac{\partial \mathbf{N}}{\partial \eta}) d\xi\, d\eta,
\end{aligned}$$

where J^* is the adjoint of J, or the inverse of J multiplied by $|J|$. For the mass matrix, all the quantities being integrated are also polynomials. Thus, all the element integrals can be done analytically, resulting in a significant savings in CPU effort over numerical integration.

As derived, Eq. (4.9) gives a coupled set of ODE's to solve for the nodal values of the state vector. The mass matrix M is sparse, positive definite (for the cases in this thesis), but unstructured, so that its inversion requires considerable computational effort. If one is only interested in the steady state, M can be replaced by a "lumped" (diagonal) matrix M_L, where each diagonal entry is the sum of all the elements in the corresponding row of M. This allows Eq. 4.9 to be solved explicitly. If one is interested in the unsteady Euler equations, it is better to invert the mass matrix with a few iterations of a preconditioned conjugate-gradient solver [44].

4.3 Choice of Test Functions

This section describes the selection of the test functions $\tilde{\mathbf{N}}$, and discusses the methods that result from each choice. To give some feel for the different kinds of test functions used, Fig. 4.1 shows perspective surfaces for the the three methods discussed below. In this figure, the heavy black line represents the element, and the height of the surface above the element represents the value of the test function. In all cases, the test function for the far right node is shown (node 3 in Fig. 3.4). Note especially that, unlike the interpolation functions, the test functions need not be zero at all other nodes. Some conditions do exist on these functions, and will be discussed in section 4.7. In two dimensions, the differences between these methods will be discussed in section 5.2 and chapter 7. For the biquadratic elements, only the Galerkin method was implemented, and in three dimensions, only the cell-vertex method was implemented. Other choices for the test functions are possible. Prozan [57] has shown how to derive upwind methods and the MacCormack method [45], and Murphy [49] has shown how several other methods fit into a finite element framework.

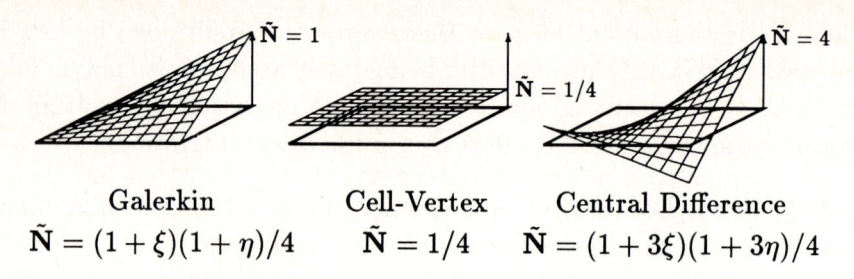

Galerkin	Cell-Vertex	Central Difference
$\tilde{N} = (1+\xi)(1+\eta)/4$	$\tilde{N} = 1/4$	$\tilde{N} = (1+3\xi)(1+3\eta)/4$

Figure 4.1: Illustrative Test Functions for Three Methods

4.3.1 Test Functions for Galerkin Method

If one choses each $\tilde{N}_i^{(e)}$ to be the corresponding $N_i^{(e)}$, one obtains the Galerkin Finite Element approximation, applied to the Euler equations in [10, 49, 55, 64]. This approximation has two interesting features. First, it gives the minimum steady-state error (in an energy norm), since there is no component of error in the space of the interpolation functions. Second, for the steady Euler equations on a uniform mesh of bilinear elements, it is a fourth-order accurate approximation. The discussion of these features is postponed until section 7.2.1. For reference, the elemental consistent mass matrix M is

$$M = \frac{1}{9}\begin{bmatrix} 4Q - 2Q_2 - 2Q_3 & 2Q - Q_3 & Q & 2Q - Q_2 \\ 2Q - Q_3 & 4Q + 2Q_2 - 2Q_3 & 2Q + Q_2 & Q \\ Q & 2Q + Q_2 & 4Q + 2Q_2 + 2Q_3 & 2Q + Q_3 \\ 2Q - Q_2 & Q & 2Q + Q_3 & 4Q + 2Q_2 + 2Q_3 \end{bmatrix}, \quad (4.11)$$

where Q, Q_2 and Q_3 are

$$Q = a_2 b_3 - a_3 b_2, \qquad (4.12a)$$
$$Q_2 = a_2 b_5 - a_5 b_2, \qquad (4.12b)$$
$$Q_3 = a_5 b_3 - a_3 b_5, \qquad (4.12c)$$

and the a's and b's are from Eq. (3.11). It is possible to assign a geometric interpretation to each of these quantities, in terms of cross products of element sides. That is, one can also write:

$$Q = \frac{1}{8}\vec{1,3} \times \vec{2,4}, \qquad (4.13a)$$
$$Q_2 = \frac{1}{8}\vec{1,2} \times \vec{4,3}, \qquad (4.13b)$$

$$Q_3 = \frac{1}{8}\vec{2,3} \times \vec{1,4}, \qquad (4.13c)$$

where the notation $\vec{i,j}$ means the vector from node i to node j. Note that Q is 1/8 of the cross product of the diagonals, so it is 1/4 of the element area. Also note that if the element is a parallelogram, then Q_2 and Q_3 will be zero. This results in the following lumped mass matrix (shown as a column vector):

$$M_L = \begin{bmatrix} Q - \frac{Q_2}{3} - \frac{Q_3}{3} \\ Q + \frac{Q_2}{3} - \frac{Q_3}{3} \\ Q + \frac{Q_2}{3} + \frac{Q_3}{3} \\ Q - \frac{Q_2}{3} + \frac{Q_3}{3} \end{bmatrix}. \qquad (4.14)$$

Also for reference, the x derivative residual matrix is

$$R_x = \frac{1}{6} \begin{bmatrix} 2(b_2 - b_3) & b_2 + 2b_3 - b_5 & b_3 - b_2 & -2b_2 - b_3 + b_5 \\ b_2 - 2b_3 - b_5 & 2(b_3 + b_2) & -2b_2 + b_3 + b_5 & -b_2 - b_3 \\ b_2 - b_3 & 2b_2 + b_3 + b_5 & 2(b_3 - b_2) & -b_2 - 2b_3 - b_5 \\ 2b_2 - b_3 + b_5 & b_2 + b_3 & -b_2 + 2b_3 - b_5 & -2(b_2 + b_3) \end{bmatrix}. \qquad (4.15)$$

A careful examination of Eqs. (4.15), (4.17), and (4.24) shows that

$$R_{\text{Galerkin}} = \frac{2}{3}R_{\text{cell-vertex}} + \frac{1}{3}R_{\text{central difference}}. \qquad (4.16)$$

This suggests that it might be possible to derive other schemes as combinations of the central difference and cell-vertex schemes, although the analysis in chapter 7 indicate that the Galerkin method results in a higher order of accuracy (for the Euler equations) than either the central difference or cell-vertex methods alone. The Galerkin method is the only method implemented for the biquadratic elements.

4.3.2 Test Functions for Cell-Vertex Method

If one chooses each $\tilde{N}_i^{(e)}$ to be a constant (in this case 1/4), for bilinear shape functions one obtains the *cell-vertex finite volume* approximation [16, 26, 56]. In two dimensions, the residual matrices are identical to the equivalent residual matrices produced by a node-based finite volume method. This will be demonstrated for the x derivatives, and the proof is identical for the y derivatives. Refer to Fig. 3.4 for a picture of the element under discussion.

The residual matrix for the x derivative is

$$R_x = \frac{1}{4}\begin{bmatrix} b_2 - b_3 & b_2 + b_3 & b_3 - b_2 & -b_3 - b_2 \\ b_2 - b_3 & b_2 + b_3 & b_3 - b_2 & -b_3 - b_2 \\ b_2 - b_3 & b_2 + b_3 & b_3 - b_2 & -b_3 - b_2 \\ b_2 - b_3 & b_2 + b_3 & b_3 - b_2 & -b_3 - b_2 \end{bmatrix} \qquad (4.17)$$

or

$$R_x = \frac{1}{2}\begin{bmatrix} y_2 - y_4 & y_3 - y_1 & y_4 - y_2 & y_1 - y_3 \\ y_2 - y_4 & y_3 - y_1 & y_4 - y_2 & y_1 - y_3 \\ y_2 - y_4 & y_3 - y_1 & y_4 - y_2 & y_1 - y_3 \\ y_2 - y_4 & y_3 - y_1 & y_4 - y_2 & y_1 - y_3 \end{bmatrix}, \qquad (4.18)$$

so the contribution of the element to all of its nodes is just

$$R_x F = \frac{1}{2}\left[(y_2 - y_4)(F_1 - F_3) + (y_3 - y_1)(F_2 - F_4)\right], \qquad (4.19)$$

where F_i and y_i are the flux vector and y coordinate at node i, and the b's are from Eq. (3.11). The line integral around the cell from the finite volume method is:

$$\oint F\,dy = \frac{F_1 + F_2}{2}(y_2 - y_1) + \frac{F_2 + F_3}{2}(y_3 - y_2) + \qquad (4.20)$$
$$+ \frac{F_3 + F_4}{2}(y_4 - y_3) + \frac{F_4 + F_1}{2}(y_1 - y_4)$$
$$= \tfrac{1}{2}\left[(y_2 - y_4)(F_1 - F_3) + (y_3 - y_1)(F_2 - F_4)\right],$$

which is identical to the result produced by the finite element approximation.

The consistent mass matrix for the cell-vertex approximation is

$$M = \frac{1}{4}\begin{bmatrix} Q - \frac{Q_2}{3} - \frac{Q_3}{3} & Q + \frac{Q_2}{3} - \frac{Q_3}{3} & Q + \frac{Q_2}{3} + \frac{Q_3}{3} & Q - \frac{Q_2}{3} + \frac{Q_3}{3} \\ Q - \frac{Q_2}{3} - \frac{Q_3}{3} & Q + \frac{Q_2}{3} - \frac{Q_3}{3} & Q + \frac{Q_2}{3} + \frac{Q_3}{3} & Q - \frac{Q_2}{3} + \frac{Q_3}{3} \\ Q - \frac{Q_2}{3} - \frac{Q_3}{3} & Q + \frac{Q_2}{3} - \frac{Q_3}{3} & Q + \frac{Q_2}{3} + \frac{Q_3}{3} & Q - \frac{Q_2}{3} + \frac{Q_3}{3} \\ Q - \frac{Q_2}{3} - \frac{Q_3}{3} & Q + \frac{Q_2}{3} - \frac{Q_3}{3} & Q + \frac{Q_2}{3} + \frac{Q_3}{3} & Q - \frac{Q_2}{3} + \frac{Q_3}{3} \end{bmatrix}, \qquad (4.21)$$

where Q, Q_2 and Q_3 are from Eq. (4.12), so the lumped mass matrix M_L is a matrix with Q along the diagonal. Since Q is one-quarter of the cell area, this results in an approximation which is identical to the node-based finite volume schemes.

In three dimensions, there may be differences from a node-based finite volume scheme, because of the way surface integrals are calculated in the finite-volume method. Many researchers assume the four nodes on a face make up a coplanar surface, but the finite element method integrates the curved surface exactly. Other properties of the cell-vertex approximation are discussed in Section 7.2.2.

4.3.3 Test Functions for Central Difference Method

If the $\tilde{N}_i^{(e)}$ are chosen to be

$$\tilde{N} = \left[\frac{(1-3\xi)(1-3\eta)}{4} , \frac{(1+3\xi)(1-3\eta)}{4} , \right. \tag{4.22}$$
$$\left. \frac{(1+3\xi)(1+3\eta)}{4} , \frac{(1-3\xi)(1+3\eta)}{4} \right],$$

one obtains the central difference or collocation approximation [57]. This approximation can also be obtained by setting $\tilde{N}$ to a series of Dirac delta functions, but this prevents the application of the conservation proof of section 4.7. For the bilinear elements on a mesh of parallelograms, this method gives the same spatial derivative as a cell-based finite volume method [28, 60]. Figure 4.2 shows a combined cell/node grid for use with the following proof. The node-based finite element scheme works with the mesh of dashed elements, and the cell-based scheme works with the solid elements. Using the finite volume approach, the x derivative at point A is calculated by a line integral around the cell:

$$(\text{Area})\frac{dF}{dx} = \tfrac{1}{2}\left[(F_A + F_B)(y_3 - y_2) + (F_A + F_D)(y_4 - y_3) + \right.$$
$$\left. + (F_A + F_F)(y_1 - y_4) + (F_A + F_H)(y_2 - y_1)\right]$$
$$= \tfrac{1}{2}\left[F_B(y_3 - y_2) + F_D(y_4 - y_3) + F_F(y_1 - y_4) + F_H(y_2 - y_1)\right]. \tag{4.23}$$

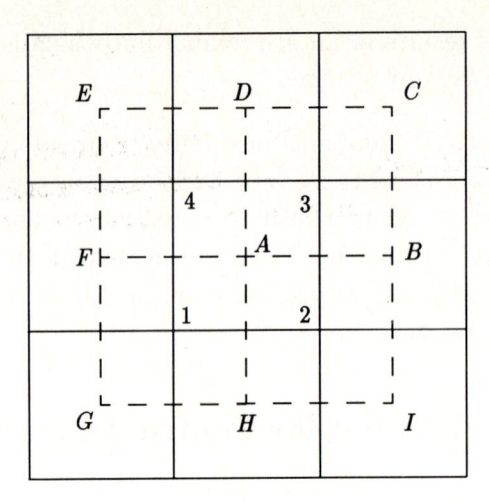

Figure 4.2: Mesh for Central Difference/Cell-Based Finite Volume Comparison

The central difference finite element method results in the following residual matrix for the x derivative:

$$R_x = \frac{1}{2} \begin{bmatrix} b_2 - b_3 & b_3 - b_5 & 0 & -b_2 + b_5 \\ -b_3 - b_5 & b_3 + b_2 & -b_2 + b_5 & 0 \\ 0 & b_2 + b_5 & b_3 - b_2 & -b_3 - b - 5 \\ b_2 + b_5 & 0 & b_3 - b_5 & -b_2 - b_3 \end{bmatrix}. \quad (4.24)$$

After applying this operator to the four elements surrounding node A, one obtains:

$$(\text{Area})\frac{dF}{dx} = \tfrac{1}{4}[F_B(y_D - y_H) + F_D(y_F - y_B) + F_F(y_H - y_D) + F_H(y_B - y_F)] \quad (4.25)$$
$$= \tfrac{1}{4}[(F_B - F_F)(y_D - y_H) + (F_D - F_H)(y_F - y_B)].$$

If the mesh is a uniform mesh of parallelograms, then

$$y_D - y_H = 2(y_3 - y_2) = 2(y_4 - y_1), \quad (4.26a)$$
$$y_F - y_B = 2(y_4 - y_3) = 2(y_1 - y_2), \quad (4.26b)$$

so the two methods produce exactly the same derivative stencil. On a non-uniform or non-parallelogram mesh, the two methods differ slightly, but the central difference finite element method still only makes use of nodes B, D, F and H.

For completeness, the consistent mass matrix for the central difference finite element method is

$$M = \frac{1}{3} \begin{bmatrix} 3Q - 2Q_2 - 2Q_3 & -Q_2 & 0 & -Q_3 \\ Q_2 & 3Q + 2Q_2 - 2Q_3 & -Q_3 & 0 \\ 0 & Q_3 & 3Q + 2Q_2 + 2Q_3 & Q_2 \\ Q_3 & 0 & -Q_2 & 3Q - 2Q_2 + 2Q_3 \end{bmatrix}, \quad (4.27)$$

and the lumped mass matrix diagonal entries are

$$M_L = \text{diag} \begin{bmatrix} Q - Q_2 - Q_3 & Q + Q_2 - Q_3 & Q + Q_2 + Q_3 & Q - Q_2 + Q_3 \end{bmatrix}, \quad (4.28)$$

where the Q's are from Eq. (4.12). Note that if the element is a parallelogram, Q_2 and Q_3 are both zero, so the lumped mass matrix is the same as for the cell-based finite volume method.

4.4 Boundary Conditions

4.4.1 Solid Surface Boundary Condition

At walls, the portions of the flux vectors representing convection normal to the wall are set to zero before each iteration, and flow tangency is enforced after each iteration. The equation for the fluxes is then

$$\mathbf{F}_w = \begin{bmatrix} \rho u_m \\ \rho u u_m + p \\ \rho u_m v \\ \rho u_m h \end{bmatrix}, \quad \mathbf{G}_w = \begin{bmatrix} \rho v_m \\ \rho u v_m \\ \rho v v_m + p \\ \rho v_m h \end{bmatrix}, \quad (4.29)$$

where u_m and v_m are corrected velocities such that the total convective contribution normal to the wall is 0. These velocities are the x and y components of the tangential velocity, given by

$$u_m = u(1 - n_x^2) - v n_x n_y, \quad (4.30)$$
$$v_m = v(1 - n_y^2) - u n_x n_y, \quad (4.31)$$

where n_x and n_y are the components of the unit normal at the node. This is easily derived from the vector expression

$$\vec{v}_{tan} = \vec{v} - (\vec{v} \cdot \hat{n})\hat{n}, \quad (4.32)$$

where $\hat{n}$ is the unit normal to the wall. This expression is obtained by finding the normal component of the velocity $(\vec{v} \cdot \hat{n})$, and subtracting it from the velocity vector.

At this point some discussion about the calculation of the normal vector is in order. In two dimensions, a parabola is fitted to the node and the nodes to its left and right, and the normal to this parabola is used as the normal at the node. In the case where the node is a corner node and has only one adjacent node, a simple linear fit is used instead. In three dimensions, the normal vector is simply the normalized sum of the cross products of the diagonals of all the faces containing that node.

At grid singularities (such as the leading and trailing edges of the scramjet fuel injector struts), the point is not treated as a boundary point, but the usual interior node formulation is used instead.

4.4.2 Open Boundary Condition

A one-dimensional characteristic treatment is used on the open or far-field boundary. From the inward-directed unit normal vector $\hat{n}$, the unit tangent vector $\hat{t}$ and the normal and tangential velocities u_n and u_t are calculated. The 1-D Riemann invariants (and the corresponding wave speeds, see Section 2.4.2) are

$$\text{invariants:} \begin{bmatrix} \dfrac{2a}{\gamma - 1} + u_n \\ \dfrac{2a}{\gamma - 1} - u_n \\ \dfrac{p}{\rho^\gamma} \\ u_t \end{bmatrix} = \begin{bmatrix} C_1 \\ C_2 \\ C_3 \\ C_4 \end{bmatrix}, \quad \text{speeds:} \begin{bmatrix} u_n + a \\ u_n - a \\ u_n \\ u_n \end{bmatrix}. \quad (4.33)$$

At each point on the boundary, the invariants are calculated using the solution state vector $\mathbf{U}$ and the "free stream" state vector $\mathbf{U}_\infty$. Then, a decision is made based on the sign of the corresponding wave speed (from the interior u_n) whether to use the invariant based on the current state, or the invariant based on the free stream. If the relevant wave speed is positive (in supersonic inflow, for example, all 4 characteristic speeds are positive), then the free stream value is used for that characteristic.

The invariants are transformed back into a set of primitive variables, and these primitive variables are used to calculate the fluxes at the boundary nodes for use in the residual calculation. These primitive variables are calculated as follows:

$$u_n = \frac{1}{2}(C_1 - C_2), \tag{4.34a}$$

$$a = \frac{\gamma - 1}{4}(C_1 + C_2), \tag{4.34b}$$

$$\rho = \left(\frac{a^2}{\gamma C_3}\right)^{1/(\gamma - 1)}, \tag{4.34c}$$

$$p = \frac{\rho a^2}{\gamma}, \tag{4.34d}$$

$$u_t = C_4, \tag{4.34e}$$

$$\rho e = \frac{p}{\gamma - 1} + \frac{1}{2}(u_n^2 + u_t^2), \tag{4.34f}$$

where C_1 - C_4 are the characteristic variables above. After the complete iteration, the characteristics are also used to update the state vectors at the boundary. Although this is not necessary for convergence, updating the state vectors after each iteration improves the robustness of the algorithm, especially for biquadratic elements.

In some problems with subsonic outflows, evaluating C_1 based on the free stream quantities is not a good boundary condition. For many problems, particularly problems involving choked flow, specifying exit pressure is a more physical condition. This characteristic treatment easily allows this. To set a specific exit pressure, the incoming characteristic (C_1) is set to the following value:

$$C_1 = \frac{4}{\gamma - 1}\sqrt{\frac{\gamma p_s}{\rho}\left(\frac{p}{p_s}\right)^{1/\gamma}} - C_2, \tag{4.35}$$

where p_s is the desired exit pressure and p is the pressure at the exit before applying the boundary condition. When inserted in Eq. (4.34), this results in $p = p_s$.

4.5 Smoothing

To capture shocks and stabilize the scheme, artificial viscosity needs to be added. The smoothing used consists of a fourth-difference term and a pressure-switched second-difference term, similar to that discussed by Rizzi and Eriksson [60]. Due to the unstructured nature of the grids, a Laplacian-type of second-difference is used, instead of normal and tangential or ξ and η differences.

The heart of the smoothing methods is the calculation of an elemental contribution to a second difference. Two ways were explored for doing this. The first method, suggested by Ni [51], is relatively fast, conservative and robust (it is dissipative on any element geometry), but gives a non-zero contribution to the second difference for a linear function on a non-uniform grid, resulting in first-order accuracy. The second method, proposed by Lindquist [38] based on the work of Mavriplis [47], is more expensive, non-conservative for quadrilaterals and less robust (it can be anti-dissipative if the element is not convex), but always results in zero contribution from linear functions on non-uniform grids, allowing second-order accuracy. Both methods were implemented in two dimensions for the bilinear elements, but only the first method was implemented for the three-dimensional elements and for the biquadratic elements.

4.5.1 Conservative, Low-Accuracy Second Difference

Figure 4.3 shows the contribution of a typical element to the second difference at node 1. The numbers inside the box are the node numbers, the numbers outside are the weights. The elemental contribution to a node is obtained by subtracting the value at the node from the average value in the element. The elemental contributions are summed to give the second difference at the node. The elemental contributions can also be multiplied by a scale factor before being summed to the node. That is, the contribution to the second difference at node 1 from element e is

$$V_1^{(e)} = k^{(e)} \left(\frac{U_1 + U_2 + U_3 + U_4}{4} - U_1 \right), \tag{4.36}$$

```
        1/4                    1/4
       ┌─────────────────────┐
       │ 4                 3 │
       │                     │
       │                     │
       │                     │
       │ 1                 2 │
       └─────────────────────┘
       -3/4                   1/4
```

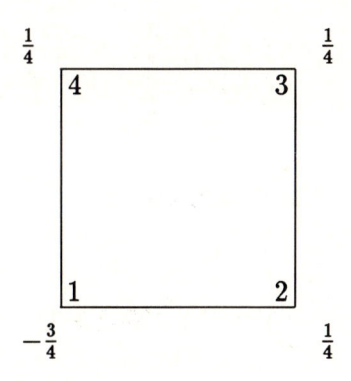

Figure 4.3: Two-Dimensional Weights for Second Difference at Node 1

where $k^{(e)}$ is some elemental weight (such as a pressure switch). This second difference method is conservative because the total contribution of each element is zero, but it is of lower accuracy since, on a non-uniform mesh, a linear function in x and y can produce a non-zero second difference at each node.

4.5.2 Non-Conservative, High-Accuracy Second Difference

This method divides the element into 4 overlapping triangles, and a line integral around each triangle is used to calculate an approximation to the first derivative at the node. Figure 4.4 shows these overlapping triangles, with the element in dashed lines and the triangle outlined in a solid line. The integration around the triangle is is used because the stencil that results from an integration around the entire quadrilateral does not damp the double-sawtooth eigenmode of the residual operator. This first derivative is integrated again around an appropriate polygon to get the second difference. The polygon and the integration direction are shown in Fig. 4.5 for a node in the interior and a node on the boundary.

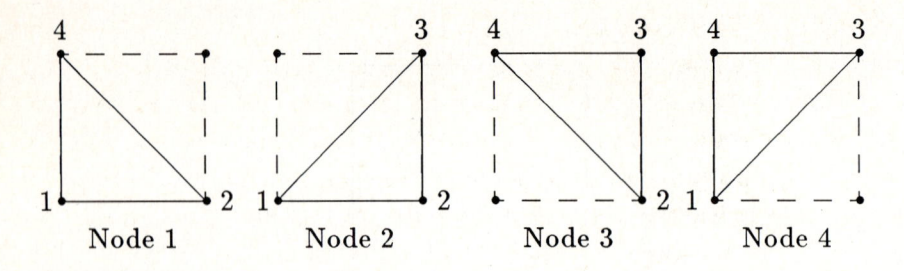

Figure 4.4: Triangles For Smoothing Calculation

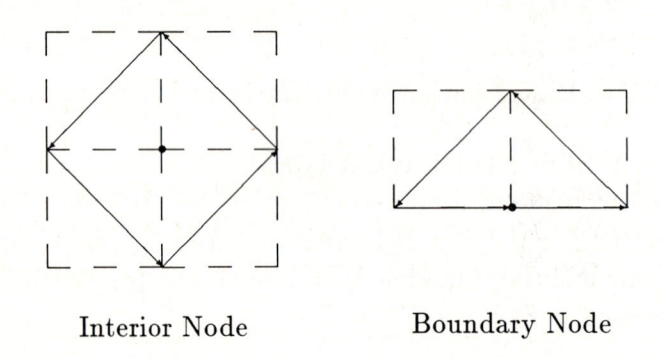

Figure 4.5: Integration Polygons for Smoothing Calculation

At node 1, for example, the contribution from an interior element is

$$U_{xx} + U_{yy} = \frac{y_4 - y_2}{2A}[U_1(y_2 - y_4) + U_2(y_4 - y_1) + U_4(y_1 - y_2)] + \frac{x_4 - x_2}{2A}[U_1(x_2 - x_4) + U_2(x_4 - x_1) + U_4(x_1 - x_2)], \quad (4.37)$$

where A is the area of triangle 1-2-4, and (x_i, y_i) are the coordinates of the ith node. For an element on the boundary, the term in front corresponding to the second integration is changed to reflect the different integration path. For example, for an element with the 1-2 face on a boundary, the terms in front would be $x_4 - x_1$ instead of $x_4 - x_2$ and $y_4 - y_1$ instead of $y_4 - y_2$, corresponding to an integration around two sides of the triangle instead of one. Note that only one factor of A is used, since a second difference is desired, not a second derivative.

4.5.3 Combined Smoothing

To calculate the complete smoothing for a time step, the nodal second difference of pressure is calculated by either of the methods described above. This is turned into an elemental quantity by simple averaging. The elemental second difference is normalized by an elemental pressure average, and this quantity is scaled so that its maximum over the entire mesh is 1. That is, the elemental pressure switch S is

$$S = \frac{\Sigma_i (\mathcal{D}_2 p)_i}{\Sigma_i p_i} \frac{1}{S_{\max}}, \qquad (4.38)$$

where $\mathcal{D}_2$ is either of the two second difference operators above, p_i is the pressure at node i, the sums are over all nodes in the element, and $S_{\max}$ is chosen so that the maximum value of S over the entire mesh is 1. The second-difference smoothing term is the weighted second difference of the state vectors (weighted by the pressure switch just described) using the conservative method above, multiplied by a constant (ν_1) between 0 and 0.05. The fourth-difference smoothing term is the second difference (by the conservative method) of the second difference (by either method) of the state vectors multiplied by a constant (ν_2) between 0.001 and 0.05. That is,

$$V_i = \nu_1 \mathcal{D}_2{}^S{}_{\text{cons}} U_i + \nu_2 \mathcal{D}_{2\text{cons}} (\mathcal{D}_2 U_i), \qquad (4.39)$$

where $\mathcal{D}_2{}^S{}_{\text{cons}}$ indicates the conservative second difference weighted by the pressure switch S, $\mathcal{D}_{2\text{cons}}$ represents the unweighted, conservative second difference operator, $\mathcal{D}_2$ represents either the high-accuracy or the low-accuracy second difference operator, and i denotes a node. The combined term V_i is added directly into the time integration of Eq. 4.40. The smoothing is globally conservative, i.e., the total contribution over the entire domain is zero. This results in first differences (when the low-accuracy method is used throughout) or third differences (when the high-accuracy method is used when applicable) normal to boundaries, but this does not affect the solutions adversely. The smoothing of section 4.5.2 is used in all test cases in the following chanpters, except where specifically noted. The high accuracy smoothing tends to be less robust for high Mach number flows, so it is typically not used in the scramjet calculations. In the current implementation, the smoothing is computed at the first stage of the multistage time integration and "frozen" for the remaining stages.

The choice of smoothing method can have a significant effect on the accuracy of a solution, and often the smoothing error is the primary source of inaccuracy in a problem. Calculations done by Lindquist and Giles [37] for the cell-vertex method and by this author for the Galerkin method indicate that, for some problems, the Galerkin and cell-vertex methods are second-order accurate when the high accuracy smoothing is used and first-order accurate when the low accuracy smoothing is used. This indicates that there is a need for further study of artificial viscosity models.

4.5.4 Smoothing on Biquadratic Elements

To calculate smoothing on the biquadratic elements, each element is subdivided into four smaller elements, and these elements are treated as bilinear elements. The smoothing method implemented in this thesis is the low accuracy smoothing of Section 4.5.1 above. Investigation of other smoothing methods for biquadratic elements is beyond the scope of the present work.

4.6 Time Integration

To integrate equation (4.9), the following multi-step method is used:

$$\begin{aligned}
U_i^{(1)} &= U_i^n + \frac{1}{4}\lambda(-\frac{\Delta t_i}{M_{Li}}R_i(U^n) + V_i^n), \\
U_i^{(2)} &= U_i^n + \frac{1}{3}\lambda(-\frac{\Delta t_i}{M_{Li}}R_i(U^{(1)}) + V_i^n), \\
U_i^{(3)} &= U_i^n + \frac{1}{2}\lambda(-\frac{\Delta t_i}{M_{Li}}R_i(U^{(2)}) + V_i^n), \\
U_i^{(4)} &= U_i^n + \lambda(-\frac{\Delta t_i}{M_{Li}}R_i(U^{(3)}) + V_i^n), \\
U_i^{n+1} &= U_i^{(4)},
\end{aligned} \qquad (4.40)$$

where $R_i(U)$ is the right-hand side of Eq. (4.9) with the fluxes based on state vector U, V_i is a smoothing term (described above), M_{Li} is the entry in the lumped mass matrix for node i, and λ is the CFL number. Local time stepping is used to accelerate convergence, with the time step given by

$$\Delta t_i = \frac{\Delta x_i}{|u| + a},\qquad(4.41)$$

where Δx_i is the minimum (over all elements containing the node) of the average lengths of opposite sides of the element, and u is the flow velocity at the node.

A Von Neumann stability analysis for the one-dimensional linear wave equation indicates that λ must be less than $2\sqrt{2}$ for stability. In two dimensions, a similar linear analysis for the wave equation $U_t + aU_x + bU_y = 0$, with $a = b = 1$, yields the following limits: λ must be less than 1.93 for the Galerkin method, less than 2.17 for the cell-vertex method, and less than 1.41 for the central difference method. This is a worst-case analysis, because if the ratio a/b is either large or small compared to 1, the Von Neumann analysis predicts a stability limit closer to the 1-D stability limit. In actual practice, the stability limits are conservative, since the Δx_i calculated is often smaller than the characteristic length limiting the stability, and since the magnitudes of the equivalents of a and b for the Euler equations (the characteristic velocities) are often quite different.

4.7 Consistency and Conservation

Consistency and conservation are two desirable properties of an Euler scheme. Consistency means that as the mesh is refined, the discrete equations approach the exact equations in some norm. Conservation means that the difference operator will not produce any spurious contributions to conserved quantities in the interior region. Conservation is important when one attempts to capture shocks or other discontinuous phenomena. This section will present (without rigorous mathematics) some sufficient conditions for consistency and conservation.

It has been shown [23] that a sufficient condition for consistency of a finite element approximation is that the element can support a constant value of the state vector (and represent a state of uniform flow) for all possible element shapes. This is equivalent to the requirement that

$$\sum N_i \equiv 1 \qquad(4.42)$$

within each element, where the sum is over all the nodes of the element.

For conservation, it is sufficient to show that the sum of each column in the assembled residual matrices R_x and R_y is zero at all interior points. This means that the contribution from each interior point is zero. The condition for conservation is that

$$\sum_i \tilde{N}_i^{(e)} = 1, \qquad (4.43)$$

where i ranges over the nodes in the element. A proof for the x derivatives is given, and the y (and z) derivatives follow similar arguments. For an interior node j, each entry in R_x is of the form

$$\int \tilde{N}_i \frac{\partial N_j}{\partial x} dV, \qquad (4.44)$$

so that the column sum is

$$\int \sum_i \tilde{N}_i \frac{\partial N_j}{\partial x} dV. \qquad (4.45)$$

Integrate Eq. (4.45) once by parts to obtain

$$\int \sum_i \tilde{N}_i \frac{\partial N_j}{\partial x} dV = \oint_{\partial V} \sum_i \tilde{N}_i N_j \hat{n} \cdot dS - \int N_j \sum_i \frac{\partial \tilde{N}_i}{\partial x} dV. \qquad (4.46)$$

The first term on the right hand side is zero because N_j is zero on the boundary (by a property of the interpolation functions), since j is an interior node. The second term will be zero if Eq. (4.43) holds, because

$$\int N_j \sum_i \frac{\partial \tilde{N}_i}{\partial x} dV = \int N_j \frac{\partial \sum_i \tilde{N}_i}{\partial x} dV \qquad (4.47)$$
$$= \int N_j \frac{\partial (1)}{\partial x} dV$$
$$= 0.$$

Thus, for interior nodes the column sum is zero, as desired. For the Galerkin, cell-vertex, and central difference approximations, $\sum \tilde{N}_i = 1$ and $\sum N_i^{(e)} = 1$ in each element, so the schemes are consistent and conservative.

4.7.1 Making Artificial Viscosity Conservative

The algorithm as presented so far introduces some slight conservation errors due to the way the artificial viscosity is added in the update scheme. This conservation error can be corrected by modifying the update step slightly. In each step of Eq. 4.40, one has

$$U_i^{(s)} = U_i^n + \alpha_s \lambda (-\frac{\Delta t_i}{M_{Li}} R_i(U^{(s-1)}) + V_i^n). \qquad (4.48)$$

To make this conservative, it is necessary to guarantee that the steady-state solution is independent of any variations in δt. To do this, one can re-arrange things slightly to obtain a modified update step,

$$U_i^{(s)} = U_i^n - \alpha_s \lambda \frac{\Delta t_i}{M_{Li}} (R_i(U^{(s-1)}) + V_i'^n), \qquad (4.49)$$

where V_i' is computed using an elemental average value of $\delta M/\delta t$ as the elemental weight $k^{(e)}$ in Eq. 4.36. The effect of this change is illustrated in section 5.2.7.

Chapter 5
Algorithm Verification and Comparisons

5.1 Introduction

This chapter contains four sections to verify the correct implementation of the finite element method. The first section examines the accuracy of the bilinear Galerkin, cell-vertex, and central difference methods in two dimensions for four model problems: $M_\infty = 2$ flow in a converging channel with a 5° convergence angle, $M_\infty = 4$ flow in a 15° converging channel, $M_\infty = 1.4$ flow in a channel over a 4% circular arc bump, and $M_\infty = 0.68$ flow in a channel over a 10% circular arc bump. The first two cases were chosen since an exact solution can be calculated for comparison purposes, and the second two were chosen since they illustrate some interesting fluid mechanics at lower Mach numbers, and since they have become traditional test cases at the MIT Computational Fluid Dynamics Laboratory. In the first test cases, comparisons with the exact solution will be presented, demonstrating the validity of the method. In all four cases, the three methods (Galerkin, cell-vertex, central difference) will be compared to see if there is any reason to choose one over the other. It will be demonstrated that the Galerkin and cell-vertex methods are better methods than the central difference method. Finally, some numerical demonstrations of conservation will be presented.

The next section examines the effect of artificial viscosity on a model problem–the 5° channel flow. The results of the test problems examined indicate that one should use the smallest coefficient of fourth difference viscosity practical, while the second difference coefficient is not as critical.

In the third section, a comparison of the bilinear and biquadratic elements is made. The test cases used are the 5° channel flow, the $M_\infty = 1.4$, 4% bump, and $M_\infty = 0.5$ flow over a 10% cosine-squared bump. The first two test cases demonstrate the ability of the biquadratic elements to handle shocks, while the last case demonstrates the biquadratic element for smooth flows.

In the final section, two examples demonstrating the extension of the cell-vertex method to three dimensions are presented.

5.2 Verification and Comparison of Methods

This section compares and verifies the Galerkin, cell-vertex, and central difference finite element methods. Except where noted, the high-accuracy damping method was used, with a fourth difference coefficient of 0.015 and a second difference coefficient of 0.03.

5.2.1 5° Converging Channel

This test case is the flow through a channel with $M_\infty = 2$ and the bottom wall sloped at 5°. Figure 5.3 shows the geometry of the channel. This problem was computed using the three methods on a coarse and fine grid. The coarse grid is shown in Fig. 5.4, and is 40x10 elements. The fine grid is 80x20 elements. Table 5.1 shows the values of pressure, density and Mach number for the exact solution in each of the five regions of the flow. To get a picture of the overall flow, contours of Mach number for the Galerkin method are presented in Fig. 5.5 and Fig. 5.6. Note that the multiple reflections are captured clearly, and also note that the shocks pass cleanly through the downstream boundary. In this problem, the normal Mach number at the first shock is 1.127. The contours ahead of the first shock are all $M = 2$ contours, and represent noise.

Figures 5.7 and 5.8 show the surface Mach numbers for the coarse and fine grids for the Galerkin method. The dotted lines represent the exact solution from oblique shock theory. Note that in the regions in between the shocks, the computed solution lies on top of the exact solution.

The next set of data presents the values of density on a slice through the grid at $y = 0.6$, for all three methods. Here, y represents the ordinary Cartesian y coordinate, with the upper wall at $y = 1$. Figure 5.9 shows the density for the coarse grid using all three methods, and Fig. 5.10

Table 5.1: Exact Solution for 5° Channel Flow

Region	Pressure	Density	Mach Number
I	.714	1.000	2.000
II	.940	1.216	1.821
III	1.218	1.463	1.649
IV	1.563	1.747	1.478
V	1.998	2.081	1.302

Table 5.2: Summary of Computed Solutions to 5° Wedge Problem

Grid	Method	% Error in ρ	% Error in p	Iterations	CPU (seconds)
40x10	Galerkin	1.9	2.6	113	19
	Cell-Vertex	1.9	2.7	117	16
	Cent. Diff.	2.0	2.9	139	25
80x20	Galerkin	1.1	1.5	197	126
	Cell-Vertex	1.1	1.5	197	101
	Cent. Diff.	1.1	1.6	240	161

shows the same results for the fine grid. Again, the dotted line represents the exact solution. Note that all three methods give essentially the same answer for this problem. This similarity occurs for all the test problems computed in this chapter. Figures 5.11 and 5.12 show the contours of constant percentage error in density for the cell-vertex method, with the lowest contour level plotted the 1% contour. This shows that over most of the field, the solution is accurate to within 1%, and that most of the remaining error is due to the finite width of the shocks. Figures 5.13 and 5.14 show all the points where the density error is greater than 5%. This shows again that the errors are mainly confined to the shocks. Table 5.2 shows the average percent error in pressure and density, the number of iteration required for convergence, and the total CPU time on a 3-processor, Alliant FX/8 (in seconds) for all three methods and both grid sizes. Note that the error is approximately halved as the grid is refined. The scheme should be second-order, but only in smooth regions of the flow, and since the error here is dominated by errors at the shocks, it is only first-order. The CPU time and iteration count comparisons will be discussed further in Section 5.2.6.

Table 5.3: Exact Solution for 15° Channel Flow

Region	Pressure	Density	Mach Number
I	.714	1.000	4.000
II	2.641	2.391	2.929
III	7.309	4.801	2.203

5.2.2 15° Converging Channel

This test case is the $M_\infty = 4$ flow through a channel with the bottom wall sloped at 15°. Figure 5.15 shows the geometry of the channel. This problem was computed using the three methods on a coarse and fine grid, and required the use of larger artficial viscosity coefficients in order to obtain convergence with the central difference method. The coarse grid is shown in Fig. 5.16, and is 33 elements by 12 elements. The fine grid is 66 elements by 24 elements. In order to correctly calculate this flow, some elements must be placed upstream of the wedge. If these upstream elements are not present, the flow quantities on the lower surface are incorrectly calculated. This is due to the treatment of the leading edge point. At all inflow boundaries, the flow is forced to be parallel to the x axis, and this results in an error which is proportional to the actual flow angle at the inflow. To get a picture of the overall flow, contours of Mach number for the Galerkin method are presented in Fig. 5.17. Table 5.3 shows the values of pressure, density and Mach number for the exact solution in each of the three regions of the flow. In this problem, the normal Mach number at the first shock is 1.82.

Figures 5.18 and 5.19 show the surface Mach numbers for the coarse and fine grids for the cell-vertex method. The dotted lines represent the exact solution from oblique shock theory. Note that there is a small error on the lower surface, but that the upper surface is correct. The next set of data presents the values of density on a slice through the grid at $y = 0.9$, for all three methods. Figure 5.20 shows the density for the coarse grid using all three methods, and Fig. 5.21 shows the same results for the fine grid. Again, the dotted line represents the exact solution. Note that all methods again give essentially the same answer. Figures 5.22 and 5.23 show all the points where the density percent error

Table 5.4: Summary of Computed Solutions to 15° Wedge Problem

Grid	Method	% Error in ρ	% Error in p	Iterations	CPU (seconds)
33x12	Galerkin	5.7	7.9	138	24
	Cell-Vertex	5.7	7.9	127	17
	Cent. Diff.	6.2	8.8	249	43
66x24	Galerkin	3.2	4.2	204	130
	Cell-Vertex	3.2	4.1	202	104
	Cent. Diff.	3.4	4.6	327	222

is greater than 5% for the cell-vertex method. This shows that the errors are mainly confined to the shocks. Table 5.4 shows the average percent error in pressure and density, the number of iteration required for convergence, and the total CPU time on a 3-processor, Alliant FX/8 (in seconds). Note the first-order behavior characteristic of shock-dominated flows. This geometry also provides a test example demonstrating a robustness problem for the central difference method. If the artificial viscosity is reduced to the values used for the test cases, the central difference method does not converge, while the cell-vertex and Galerkin methods do not have any particular problem. This is apparently due to errors near the shock reflection, which result in a pseudo-unsteadiness to the problem. Figure 5.1 shows the comparative convergence histories for the three methods for this failed case. All methods used the same CFL number (equal to 2) and artificial smoothing parameters.

5.2.3 4% Circular Arc Bump

This test case is $M_\infty = 1.4$ flow over a 4% circular arc bump in a channel, calculated on a 60x20 grid. Figure 5.24 shows the pressure contours calculated by the Galerkin method, and Fig. 5.25 shows the surface Mach number for the cell-vertex method. There are two features worthy of special note. First, note the reflection of the leading edge shock, which results in a small Mach stem and subsonic region on the upper wall. Second, note the coalescence of the trailing edge shock and the twice-reflected leading edge shock. These features will be presented in more detail in Section 5.4.2. Finally, the density at mid-channel is shown for all three methods in Fig. 5.26. The differences between the

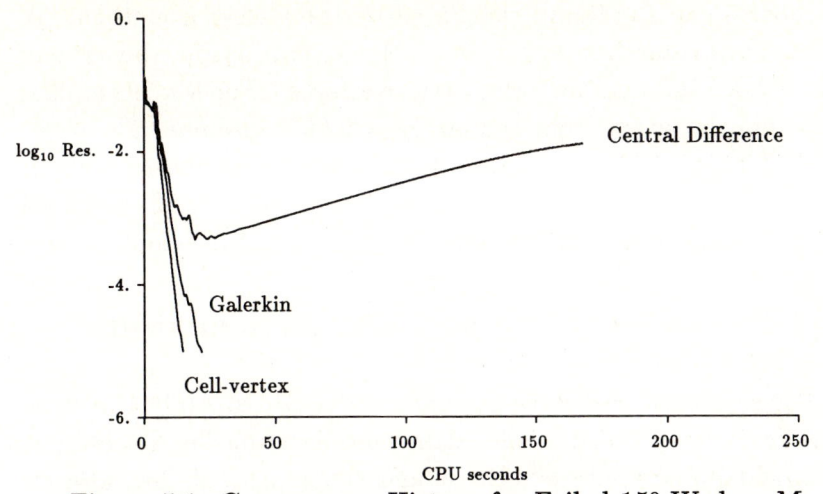

Figure 5.1: Convergence History for Failed 15° Wedge, $M_\infty = 4$

methods are not significant here.

5.2.4 10% Circular Arc Bump

This test case is the $M_\infty = 0.68$ flow over a 10% circular arc bump in a channel, calculated on a 60x20 grid. Figure 5.27 shows the pressure contours calculated by the Galerkin method. This case has a normal shock standing on the bump, with a maximum Mach number ahead of the shock of about 1.5. Figure 5.28 shows the surface Mach number for the cell-vertex method. Note that the shock is captured over 3 points, and note the slight overshoots in the shock. Finally, the density for all three methods on a slice at $y = 0.25$ is presented in Fig. 5.29. The three methods are in very close agreement. The CPU times for the three methods were quite different, however, and the next section discusses some of these differences.

5.2.5 10% Cosine Bump

This test case is the $M_\infty = 0.5$ flow over a 10% cosine-squared bump in a channel, calculated on a 60x20 grid. This flow remains subsonic throughout, so the solution should be symmetric. Figure 5.30 shows

Mach number along the bump calculated by the Galerkin method. To accentuate any asymmetries in the solution, the solutions for the left half and right half of the domain have been overlayed. For Mach number, The maximum deviation from symmetry is 0.0013, (about .2% of the average Mach number), and the RMS deviation from symmetry is about 0.1%. The error is due to spurious entropy generated by the artificial viscosity, and reductions in artificial viscosity reduce the errors.

5.2.6 CPU Comparison and Recommendations

Since the results of the three methods are approximately the same for many practical problems, the main considerations in selecting a method should be computational cost and robustness. It has already been demonstrated that the central difference method is less robust than the Galerkin or cell-vertex methods. Also, the three methods differ in their stability bounds, and in the amount of effort it takes to calculate the residuals. These effects combine to produce a marked difference in CPU work between the methods. Figure 5.31 shows the residual history plotted against CPU time for the 4% bump problem. All three methods were run at the maximum *experimentally determined* stable CFL number. For further comparison, Table 5.5 shows the CFL number, number of iterations, and total CPU time for the 4% and 10% bump problems. The codes for the Galerkin and central difference methods are virtually identical, so CPU differences are not a result of differences in programming technique. The cell-vertex method uses an optimized method of calculating residuals not available for the Galerkin and central difference methods, so its times are improved by the coding techniques. Note that the central difference method is by far the poorest, while the Galerkin method is slightly worse than the cell-vertex method. Based on these results, and on the results of the dispersion analysis in chapter 7, both the cell-vertex and Galerkin methods are recommended, while it is suggested that the central difference method be rejected.

5.2.7 Verification of Conservation

In order to verify that the schemes are conservative, the total mass flux through the channel was computed at several stations for a variety

Table 5.5: Comparison of Computational Effort for Circular Arc Bumps

Case	CFL	Iterations	CPU seconds
4% Galerkin	2.8	203	97
4% cell-vertex	2.8	204	77
4% central difference	2.2	248	126
10% Galerkin	2.7	500	239
10% cell-vertex	2.7	509	191
10% central difference	2.0	644	325

Table 5.6: Conservation in $M_\infty = 1.4$, 4% Bump Case

Case	$x = -1.4$	$x = 0$	$x = .5$	$x = 1.4$	Max. Error
30x10 Biquadratic	1.40021	1.39847	1.40071	1.40106	0.11%
60x20 Cell-Vertex	1.40000	1.40077	1.40281	1.40184	0.19%
60x20 Galerkin	1.40000	1.40095	1.40199	1.40201	0.14%
60x20 Central Diff.	1.40000	1.40138	1.40097	1.40210	0.14%

of problems. Table 5.6 shows the total mass flux through the channel at $x = -1.4$, $x = 0$, $x = 0.5$ and $x = 1.4$ for the Galerkin, cell-vertex, central difference, and biquadratic methods. For the Galerkin method, the conservative modification to the smoothing was also used as a test case. The exact mass flux should be 1.4, and the worst-case error for this problem was only 0.2%. For the 60x20, cell-vertex, $M_\infty = 0.68$, 10% bump case, the maximum error was 0.14% and for the 80x20 cell-vertex, $M_\infty = 2$, 5° wedge case, the maximum error was 0.10%. The slight errors in conservation are probably due to some combination of round-off error, local time stepping and truncation error in the algorithm used to compute the mass flux through a slice.

5.3 Effects of Added Dissipation

In order to stabilize the computational schemes, artificial dissipation is added. This dissipation was discussed more fully in section 4.5, but the essentials are repeated here. The smoothing consists of a pressure-switched second difference to capture shocks, and a fourth difference background smoothing to stabilize the scheme. Each of these smoothings has an associated coefficient. This section discusses the effects of changing these coefficients. In all the test examples shown here, the cell-vertex spatial discretization and the high-accuracy smoothing method are used. For these problems, the low-accuracy method solutions differ only slightly from the high-accuracy method solutions. The test case used is $M_\infty = 2$ flow in the 5° channel. This problem was examined with the coefficient of the second difference (ν_1) set to 0.004 and 0.04, and the coefficient of the fourth difference (ν_2) set to 0.004 and 0.02. The calculations demonstrated that the amount of second-difference smoothing is less important than the amount of fourth-difference smoothing. Figure 5.32 shows the mid-channel density in the case where $\nu_1 = 0.04$ and $\nu_2 = 0.004$. Note that the shocks are spread over 5 or 6 points, but that there is almost no high-frequency overshoot near the shocks. The spreading is still present with the low value of ν_1 (Fig. 5.33), but there is more overshoot. The spreading is due mainly to the fact that the shocks are not very strong, so they do not self-steepen (the behavior is only slightly nonlinear). Also note the small low-frequency oscillation preceding each shock. This oscillation is due to dispersion, and is explained in detail in chapter 7. Finally, in the region between the shocks, note that the solution is almost constant. Figure 5.34 shows the mid-channel density in the case where $\nu_1 = 0.04$ and $\nu_2 = 0.02$. Note that in this case, the regions between the shocks exhibit some variation, and the shocks are spread out even further.

Entropy is a more sensitive measure of the effects of dissipation than density. Figure 5.35 shows the mid-channel entropy for the case with large ν_2, and Fig. 5.36 shows the entropy for the case with small ν_2. Note that with large ν_2 there are more oscillations near the shocks. Also note that the entropy undergoes a non-physical increase between $x = 3$ and the outflow boundary. This increase is not as pronounced as Fig. 5.36. These cases indicate that the amount of the global, fourth difference smoothing is an important factor in solution quality, and that one should try to use as little fourth difference smoothing as possible.

5.4 Biquadratic *vs.* Bilinear

Biquadratic elements have the potential to produce more accurate solutions on coarser meshes. The use of coarser meshes may allow the CPU cost for a given accuracy to be reduced. On the other hand, biquadratic elements may have problems with Gibbs' phenomenon, producing unacceptable oscillations near shocks. Furthermore, a biquadratic element requires more CPU time to calculate element residuals, possibly resulting in *more* CPU effort for a given problem. A single biquadratic Galerkin element requires a factor of 3.3 more CPU time than a single bilinear Galerkin element, or about a factor of 4.2 more CPU time than a single cell-vertex element. The purpose of this section is to demonstrate that the advantages of biquadratic elements outweigh the disadvantages, resulting in a net CPU savings. Three test problems are used to compare the bilinear and biquadratic formulations of the Galerkin method in two dimensions. These problems are the 5° channel flow at Mach 2, the $M_\infty = 1.4$ flow over a 4% bump, and the flow over a 10% cosine-squared bump in a channel at Mach 0.5. The first two cases demonstrate the ability of the biquadratic elements to handle shocks. Other test cases presented later have verified that the oscillations produced by shocks are not unacceptable for a wide range of problems (see Section 8.2). The third case demonstrates the biquadratic elements in smooth flows.

5.4.1 5° Channel Flow

The 5° channel flow was computed using biquadratic elements on three different grids: 12x3, 20x5 and 40x10. The surface densities for the coarsest and finest meshes are shown in Figs. 5.37 and 5.38. Note the sharpness of the shocks and the lack of oscillations. Also note the near absence of low frequency dispersive error. The comparison with the exact solution is excellent. Figures 5.39 and 5.40 show the points where the density error exceeds 5%. Note that the result for the 12x3 mesh compares favorably with the bilinear 40x10 mesh, and the 40x10 biquadratic mesh is comparable to the bilinear 80x20 mesh. Table 5.7 shows a comparison of average error and computational effort for these cases. This table shows some interesting facts. For about half the effort of the 40x10 bilinear case, one can obtain the same accuracy using

Table 5.7: Comparison of Bilinear and Biquadratic Galerkin Solutions to 5° Wedge Problem

Case	% Error in ρ	% Error in p	Iterations	CPU (seconds)
40x10 Bilinear	1.9	2.6	113	19
80x20 Bilinear	1.1	1.5	197	126
12x3 Biquadratic	2.1	2.9	125	9
20x5 Biquadratic	1.3	1.9	170	26
40x10 Biquadratic	0.7	1.0	297	150

a 12x3 biquadratic mesh. For about the same effort as an 80x20 bilinear mesh, a 40x10 biquadratic mesh provides better accuracy.

5.4.2 4% Circular Arc Bump

The 4% bump presented above was computed using biquadratic elements on a 30x10 mesh. Pressure contours for this case are shown in Fig. 5.41, and illustrate the ability of biquadratic elements to resolve sharp gradients. Compare Fig. 5.41 with Fig. 5.24 and note how much better all the interactions are resolved. Both the 30x10 biquadratic mesh and the 60x20 bilinear mesh have 1281 nodes, but except for some noise, the solution on the biquadratic grid is closer to a solution on a 120x40 bilinear grid with 4800 elements and 4961 nodes (see Fig. 5.42) than it is to the 60x20 bilinear solution. The biquadratic case required 131 seconds on the Alliant, while the 120x40 bilinear case required 592 seconds.

5.4.3 10% Cosine Bump

One expects the biquadratic elements to be very good for smooth flows. To verify this, $M_\infty = 0.5$ flow over a 10% cosine-squared bump was computed on a 24x8 biquadratic mesh and a 60x20 bilinear mesh. Figure 5.43 shows contours of density for the biquadratic elements. The contours are quite symmetric, as one would expect from a flow which remains completely subsonic. Most of the non-smoothness seen in the contours is introduced by the plot package (which divided each biquadratic element into 32 linear triangles), rather than actual errors in the flow. For comparison, Fig. 5.44 shows these contours in the bilinear case. The agreement is quite good. A real advantage of the biquadratic elements

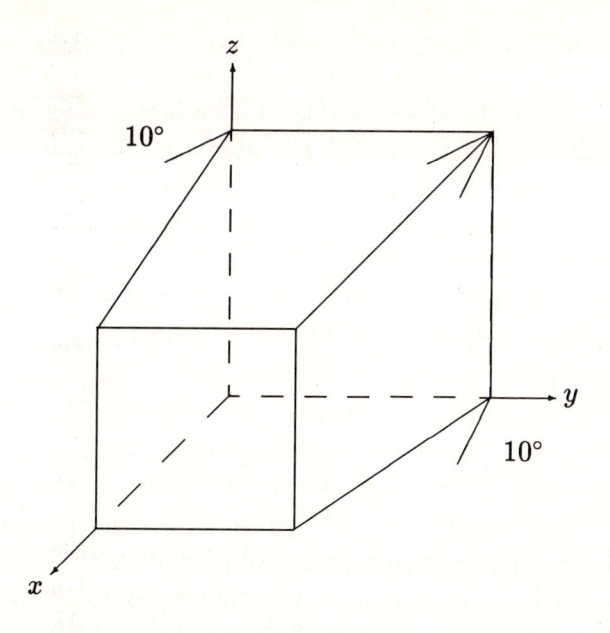

Figure 5.2: Geometry for 10° Double Wedge Flow

is the CPU savings they produce. Figure 5.45 shows the residual history versus CPU time for both bilinear and biquadratic methods. The biquadratic method converges somewhat faster, even though more iterations were actually required in this case.

These test cases demonstrate the potential utility of the higher-order element. The main area for further research before biquadratic elements can be used routinely is in the creation of better artificial dissipation formulations, in order to enable one to achieve the maximum accuracy available from these elements.

5.5 Three Dimensional Verification

To verify the three dimensional code, two test cases were computed. The first test case is quasi two-dimensional flow in a converging channel with $M_\infty = 2$. The channel has one wall sloping at a 5° angle, and all other walls are aligned with the flow. The second test case is $M_\infty = 2.5$ flow in a converging channel with both the y and z walls sloped at 10°. The geometry for this case is shown in Fig. 5.2. This figure also gives the axis orientations for the first test case.

The $M_\infty = 2$, 5° flow was computed on a 30x30x30 grid. This should produce a single shock, with shock jumps identical to those of Section 5.2.1. Figure 5.46 shows density contours on a slice through the channel perpendicular to the x axis. The curvature of the contours near the boundaries is due to artificial viscosity. As the artificial viscosity is increased, the curvature is also increased. Figure 5.47 shows the pressure contours, as well as a slice of the computational grid. The slice is taken perpendicular to the y axis, midway into the channel. Note the shock coming up from the bottom wall. Finally, Fig. 5.48 shows the Mach number on a slice perpendicular to the z axis, at $z = 0.7$. The exact shock jump values are shown in Table 5.1. In this case the Mach number after the shock was 1.81 (exact 1.82), the pressure was 0.939 (exact 0.940), and the density was 1.22 (exact 1.22). The agreement is very good, indicating the soundness of the algorithm. This case required 240 iterations to converge, and took 72 minutes on the Alliant.

The $M_\infty = 2.5$, 10° double wedge case was also computed on a 30x30x30 grid. The double wedge introduces regions in which the flow may have passed through zero, one or two shocks. In addition, there are other kinds of interactions present, and these are discussed by Kutler in [34]. In this case, Fig. 5.49 presents the density in a slice perpendicular to the flow direction. The shocks can be seen clearly, and in the region where the flow has passed through both shocks (the upper right corner), the results of the interactions are visible. In particular, note the bending of the shocks as they pass through each other and interact. This behavior is more pronounced as the shock strength increases. Figure 5.50 shows the Mach number on the plane $y = 0.5$. Here, the bending and interaction is much more pronounced. Finally, Fig. 5.51 demonstrates the same behavior with pressure on a slice at $z = 0.7$.

5.6 Summary

This chapter has demonstrated the soundness of the three algorithms in both two and three dimensions. Results verifying the conservation of mass for all three methods were presented for a variety of problems. The Galerkin and cell-vertex methods were shown to require less CPU time than the central difference method, and the central difference method was demonstrated to be less robust than the other methods. Biquadratic elements were examined and show promise.

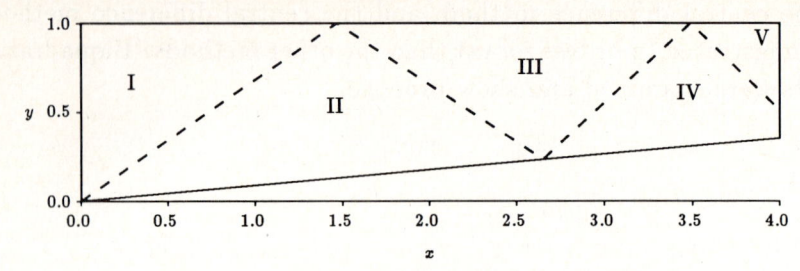

Figure 5.3: Flow Geometry for 5° Channel

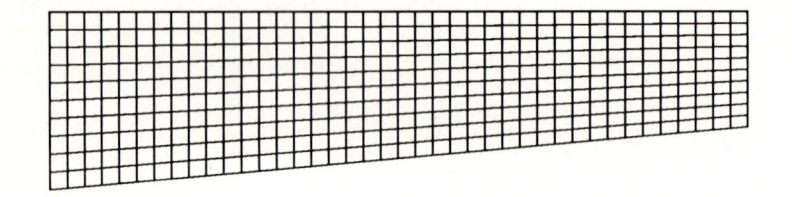

Figure 5.4: Coarse Grid, 5° Channel

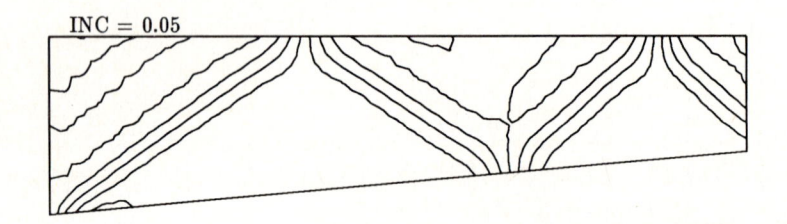

Figure 5.5: Mach Number, Galerkin, 5° Channel, M=2, 40x10 Grid

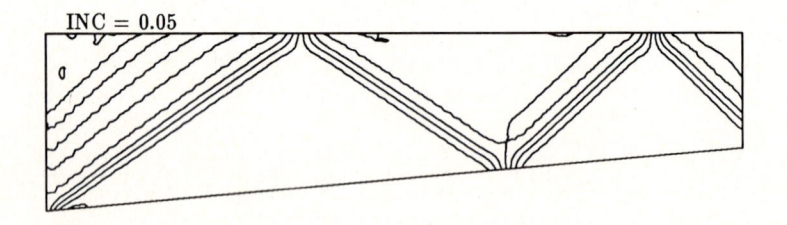

Figure 5.6: Mach Number, Galerkin, 5° Channel, M=2, 80x20 Grid

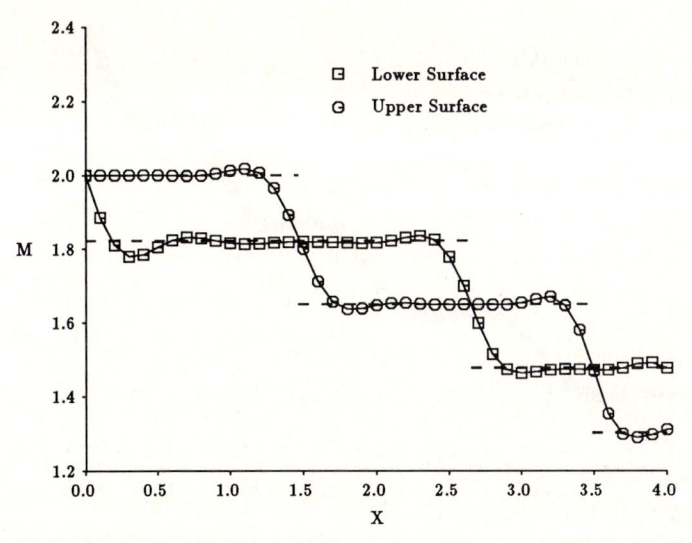

Figure 5.7: Surface Mach Number, Galerkin, 5° Channel, M=2, 40x10 Grid

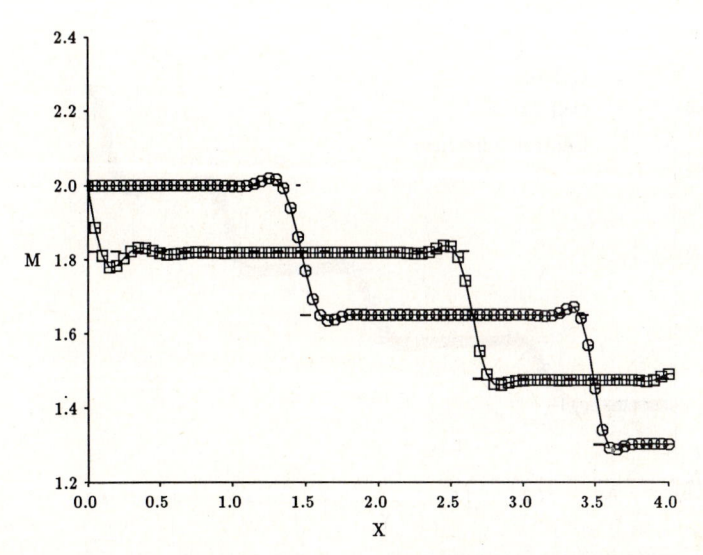

Figure 5.8: Surface Mach Number, Galerkin, 5° Channel, M=2, 80x20 Grid

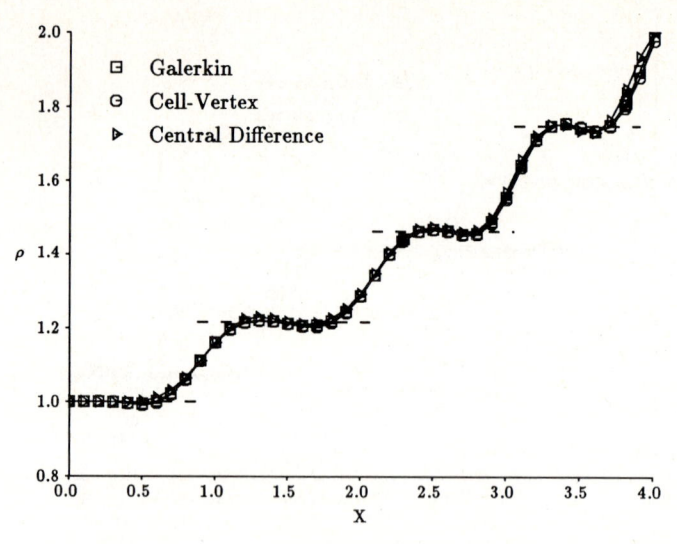

Figure 5.9: Density Comparison, 5° Channel, M=2, 40x10 Grid, Y=0.6

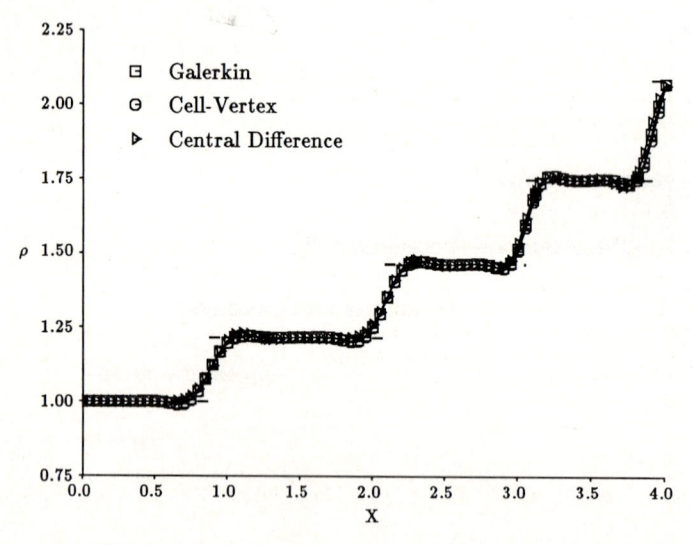

Figure 5.10: Density Comparison, 5° Channel, M=2, 80x20 Grid, Y=0.6

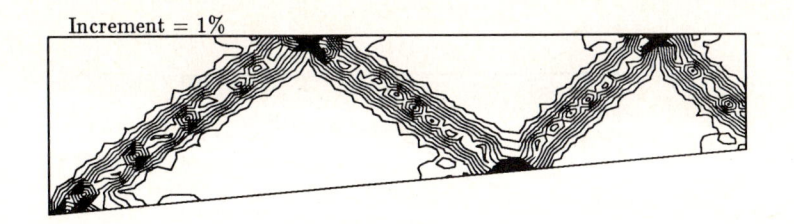

Figure 5.11: Percent Error in Density, 5° Channel, M=2, 40x10 Grid

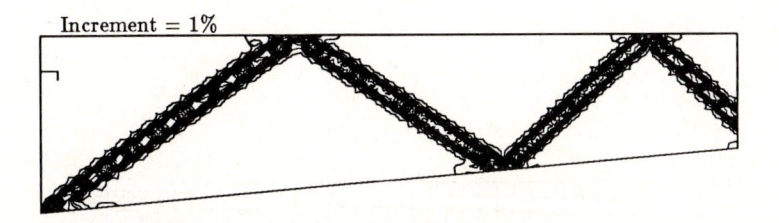

Figure 5.12: Percent Error in Density, 5° Channel, M=2, 80x20 Grid

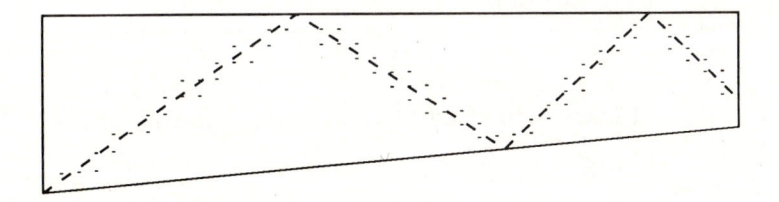

Figure 5.13: Points with Density Error > 5%, 5° Channel, M=2, 40x10 Grid

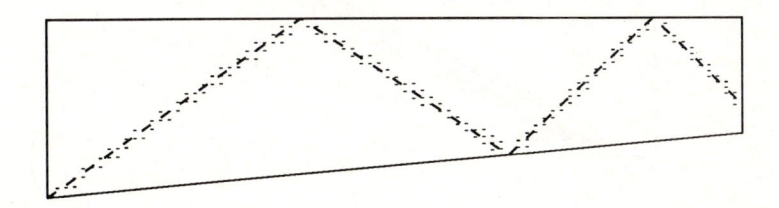

Figure 5.14: Points with Density Error > 5%, 5° Channel, M=2, 80x20 Grid

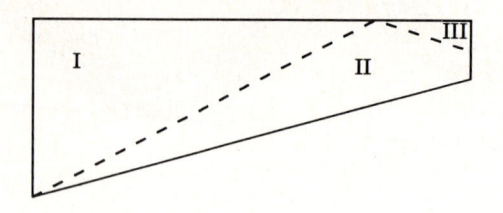

Figure 5.15: Flow Geometry for 15° Channel

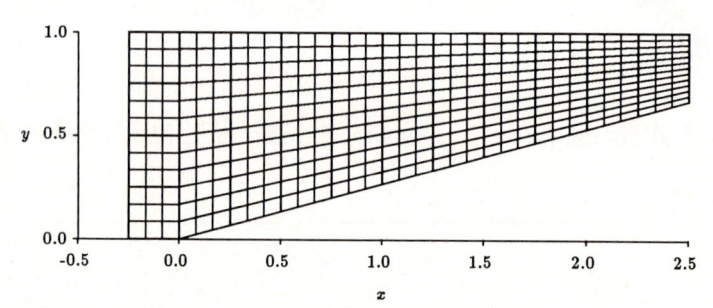

Figure 5.16: Coarse Grid, 15° Channel

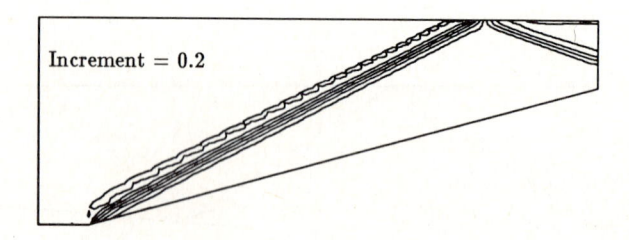

Figure 5.17: Mach Number, Galerkin, 15° Channel, M=4, 66x24 Grid

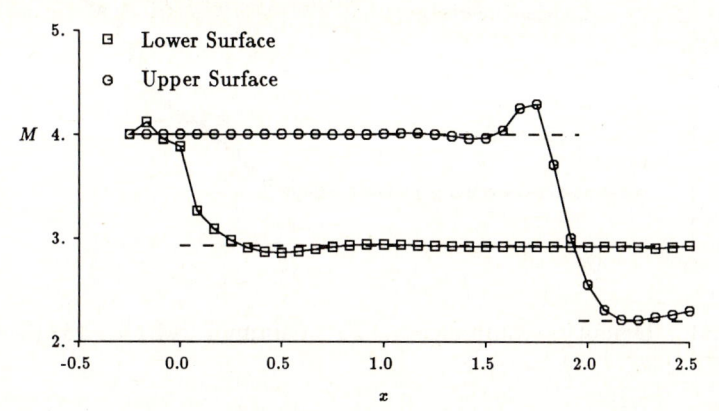

Figure 5.18: Surface Mach Number, Cell-Vertex, 15° Channel, M=4, 33x12 Grid

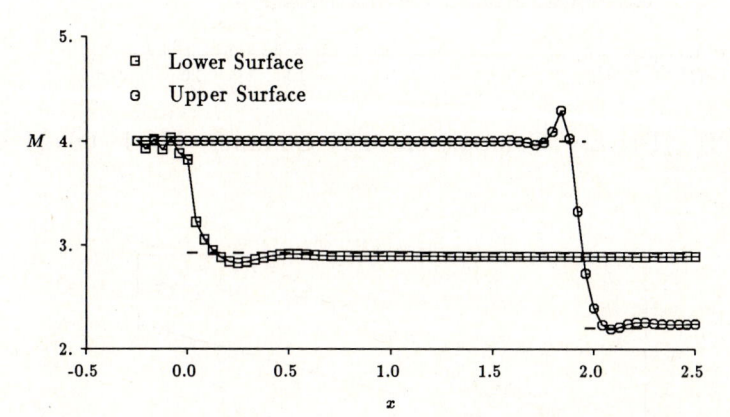

Figure 5.19: Surface Mach Number, Cell-Vertex, 15° Channel, M=4, 66x24 Grid

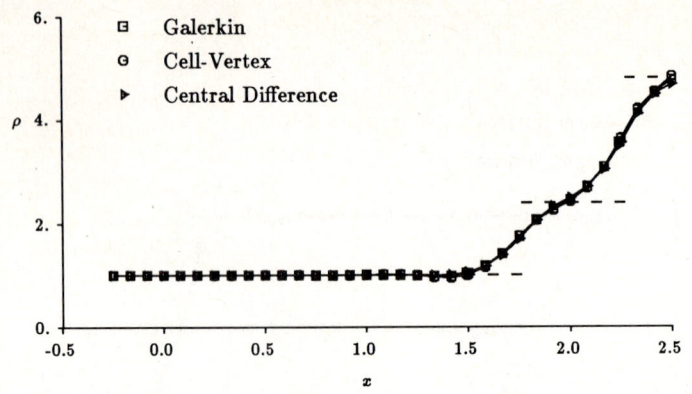

Figure 5.20: Density Comparison, 15° Channel, M=4, 33x12 Grid, Y=0.9

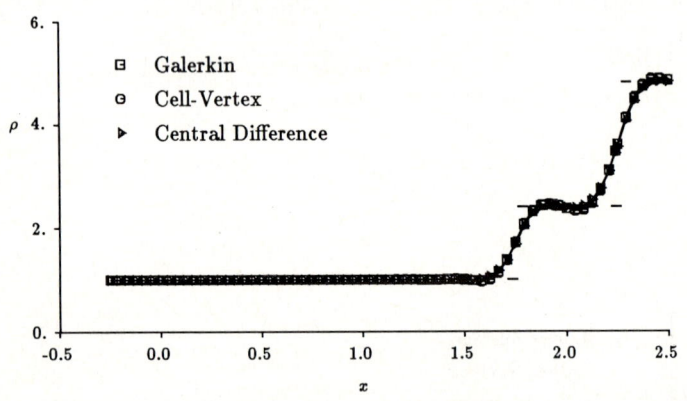

Figure 5.21: Density Comparison, 15° Channel, M=4, 66x24 Grid, Y=0.9

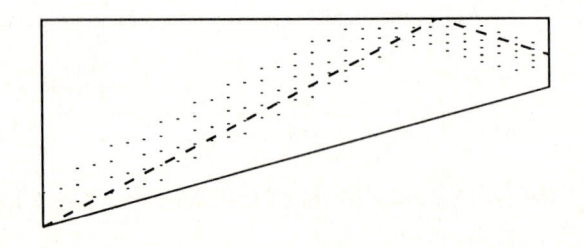

Figure 5.22: Points with Density Error > 5%, 15° Channel, M=4, 33x12 Grid

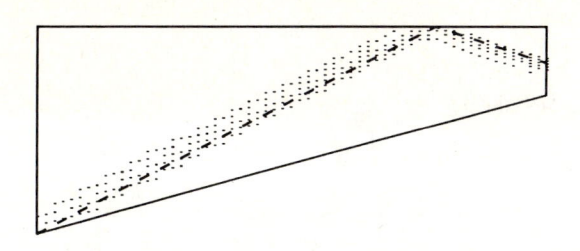

Figure 5.23: Points with Density Error > 5%, 15° Channel, M=4, 66x24 Grid

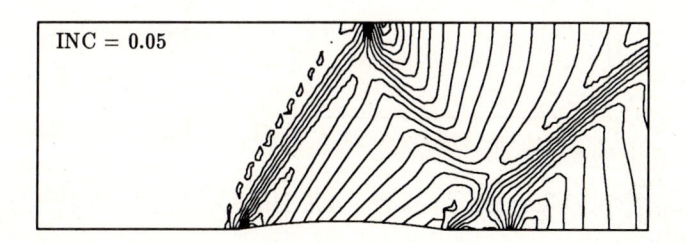

Figure 5.24: Pressure Contours, 4% Bump, $M_\infty = 1.4$, 60x20 Grid, Galerkin

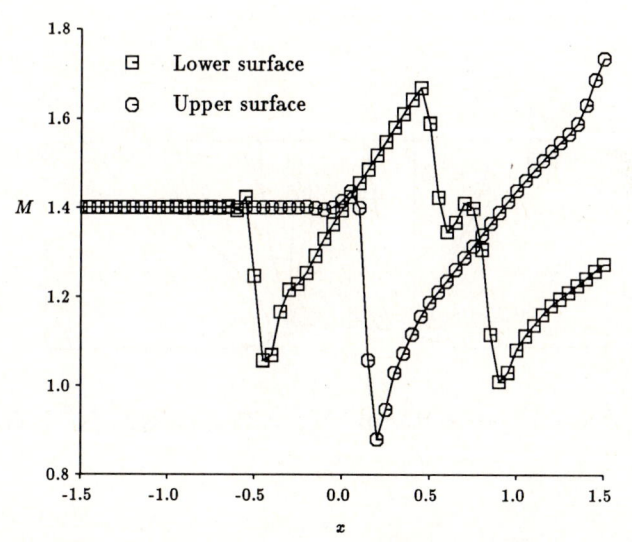

Figure 5.25: Surface Mach Number, 4% Bump, $M_\infty = 1.4$, 60x20 Grid, Cell-Vertex

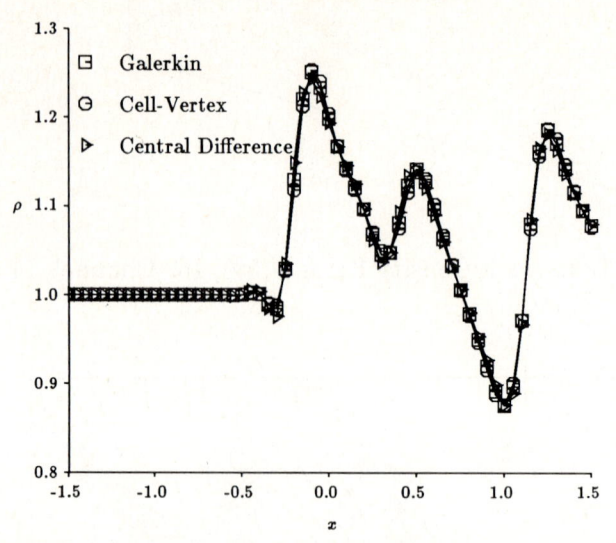

Figure 5.26: Mid-Channel Density, 4% Bump, $M_\infty = 1.4$, 60x20 Grid, All Methods

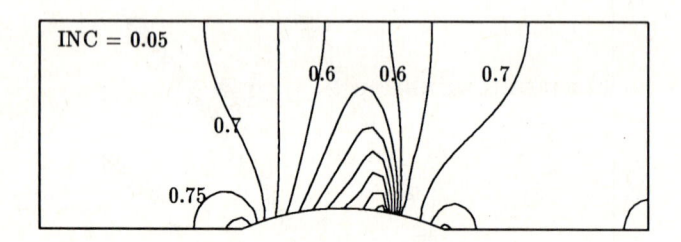

Figure 5.27: Pressure, 10% Bump, $M_\infty = 0.68$, 60x20 Grid, Galerkin

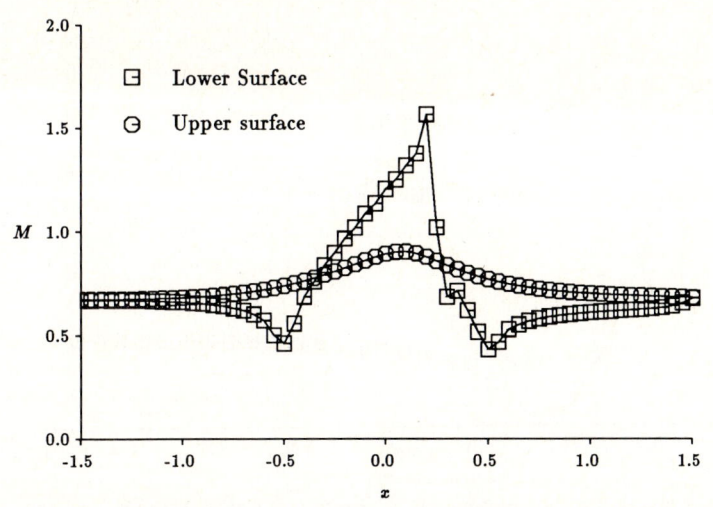

Figure 5.28: Surface Mach Number, 10% Bump, $M_\infty = 0.68$, 60x20 Grid, Cell-Vertex

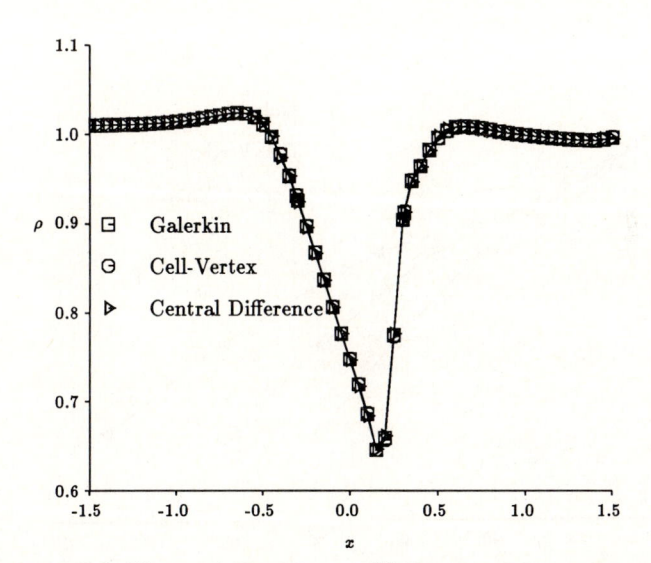

Figure 5.29: Mid-Channel Density, 10% Bump, $M_\infty = 0.68$, 60x20 Grid, All Methods

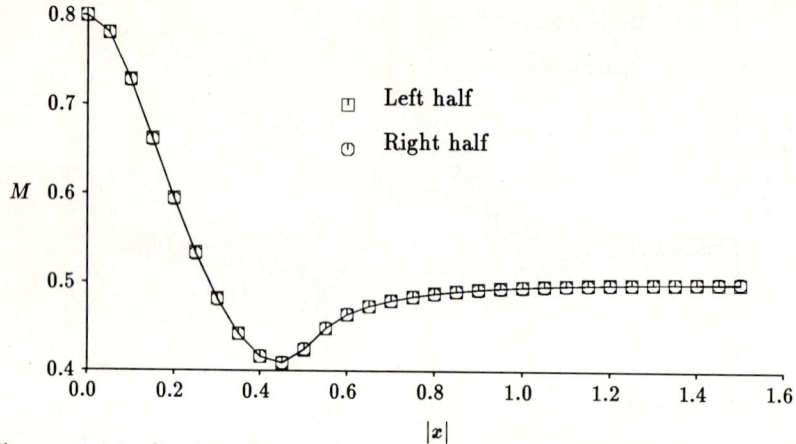

Figure 5.30: Surface Mach Number, 10% $\cos^2$ Bump, $M_\infty = 0.5$, 60x20 Grid

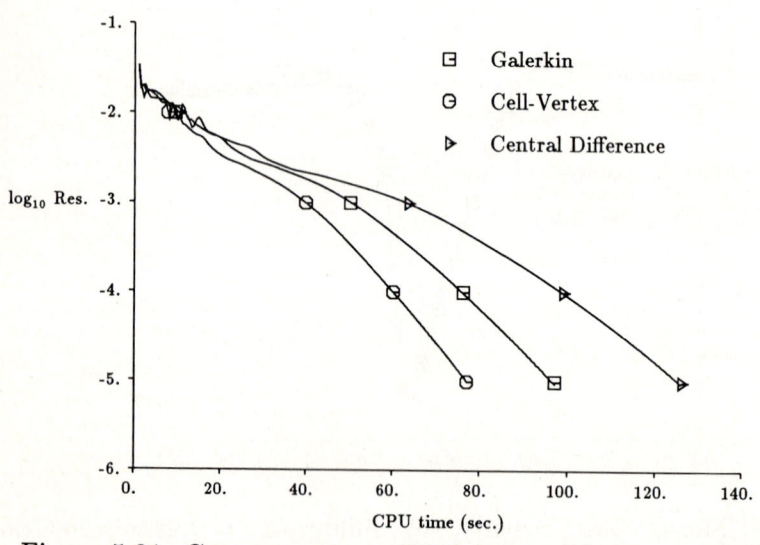

Figure 5.31: Convergence vs. CPU Time, 4% Bump, All Methods

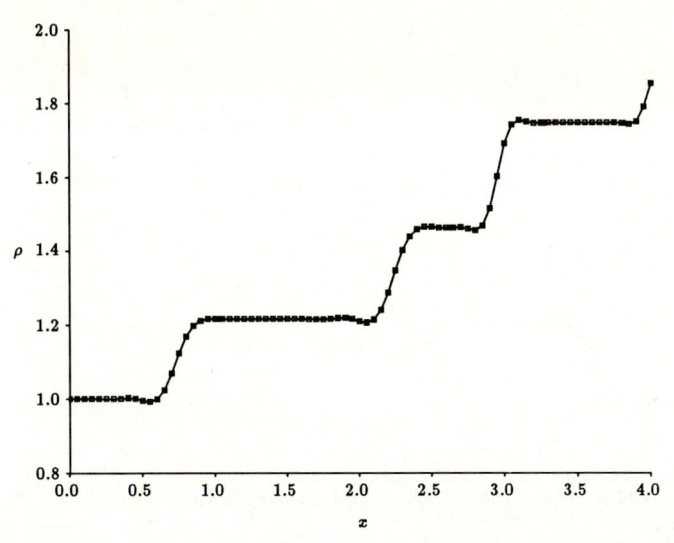

Figure 5.32: Mid-channel Density, $\nu_1 = 0.04$, $\nu_2 = 0.004$, 5° Wedge

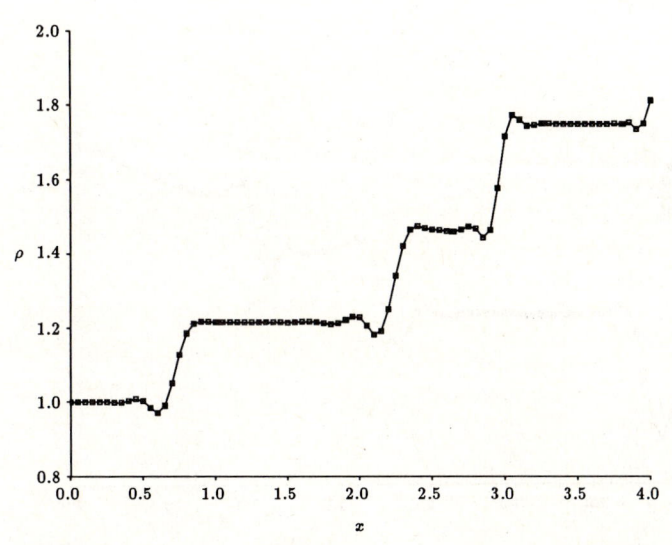

Figure 5.33: Mid-channel Density, $\nu_1 = 0.004$, $\nu_2 = 0.004$, 5° Wedge

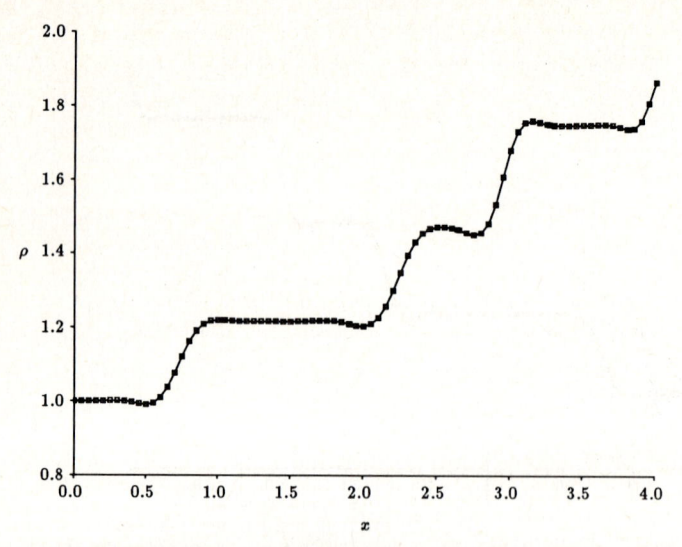

Figure 5.34: Mid-channel Density, $\nu_1 = 0.04$, $\nu_2 = 0.02$, 5° Wedge

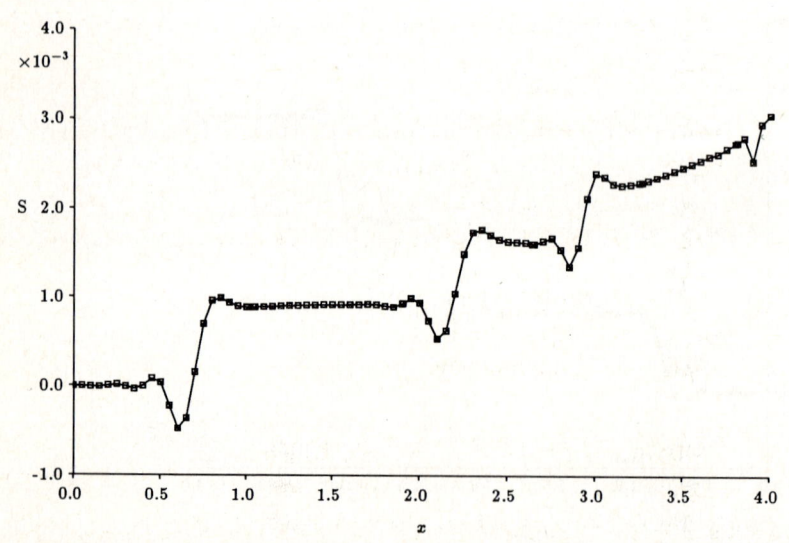

Figure 5.35: Mid-channel Entropy, $\nu_1 = 0.04$, $\nu_2 = 0.02$, 5° Wedge

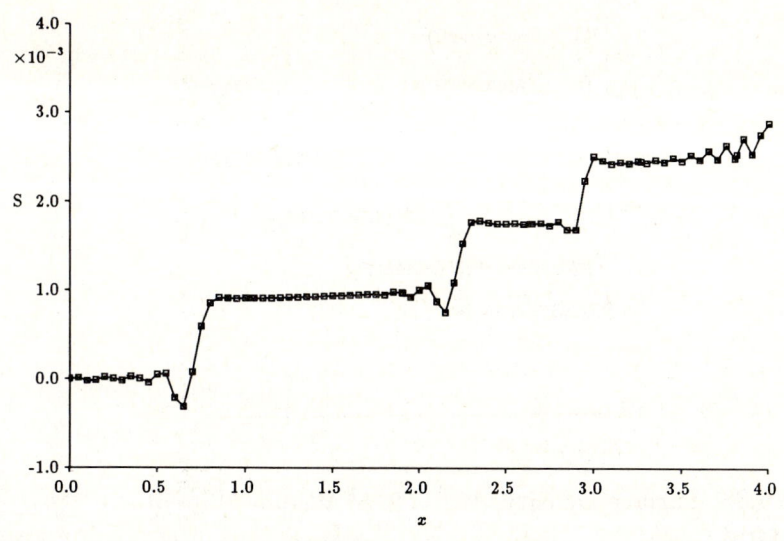

Figure 5.36: Mid-channel Entropy, $\nu_1 = 0.04$, $\nu_2 = 0.004$, 5° Wedge

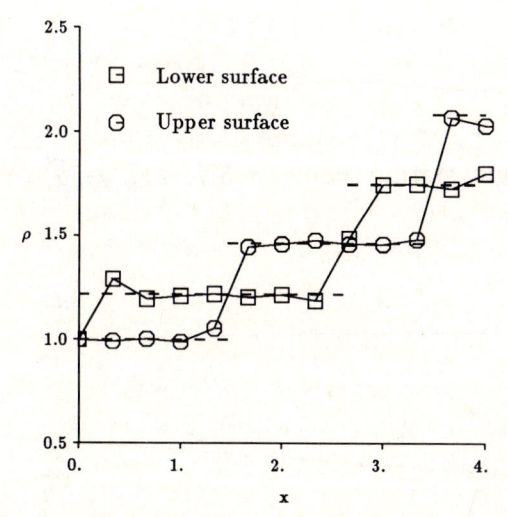

Figure 5.37: Surface Density, $M_\infty = 2$, 5° Wedge, Biquadratic Elements, 12x3 Grid

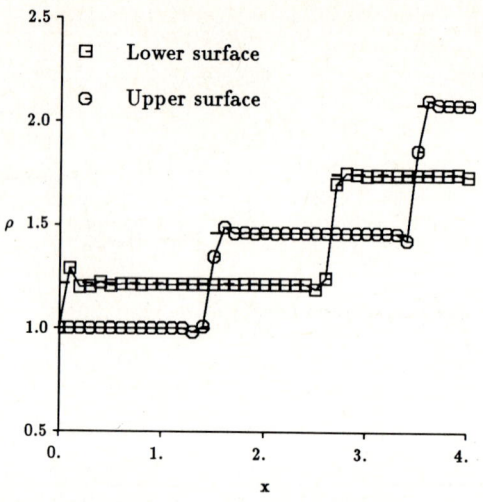

Figure 5.38: Surface Density, $M_\infty = 2$, 5° Wedge, Biquadratic Elements, 40x10 Grid

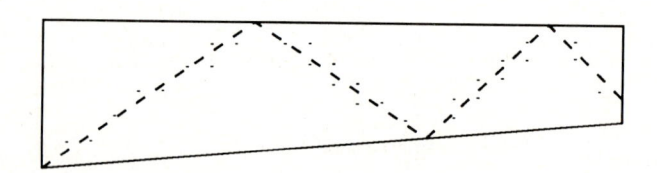

Figure 5.39: Points With ρ Error > 5%, $M_\infty = 2$, 5° Wedge, 12x3 Biquadratic Grid

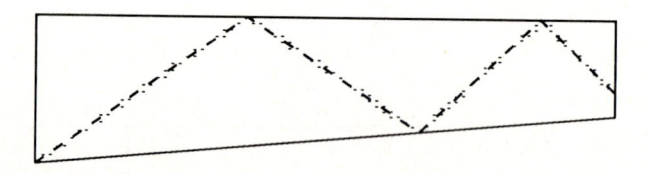

Figure 5.40: Points With ρ Error > 5%, $M_\infty = 2$, 5° Wedge, 40x10 Biquadratic Grid

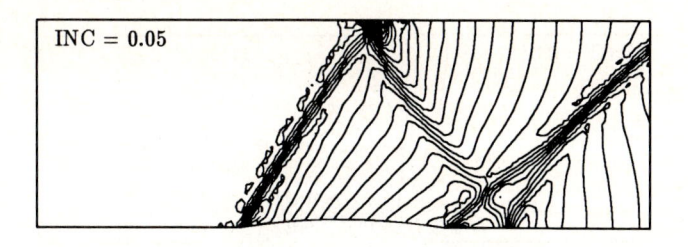

Figure 5.41: Pressure Contours, 4% Bump, $M_\infty = 1.4$, 30x10 Grid, Biquadratic Elements

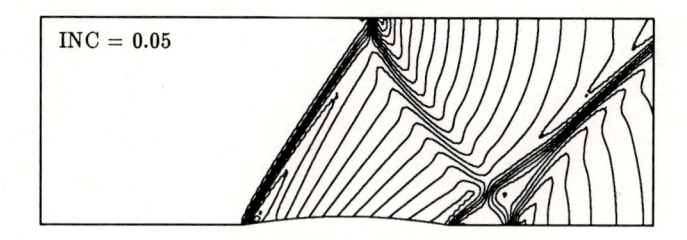

Figure 5.42: Pressure Contours, 4% Bump, $M_\infty = 1.4$, 120x40 Grid, Bilinear Elements

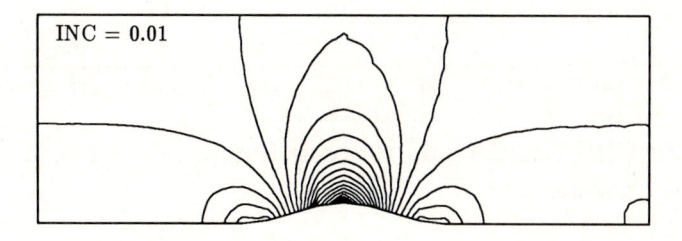

Figure 5.43: Density, 10% $\cos^2$ Bump, $M_\infty = 0.5$, 24x8 Grid, Biquadratic Elements

Figure 5.44: Density, 10% $\cos^2$ Bump, $M_\infty = 0.5$, 60x20 Grid, Bilinear Elements

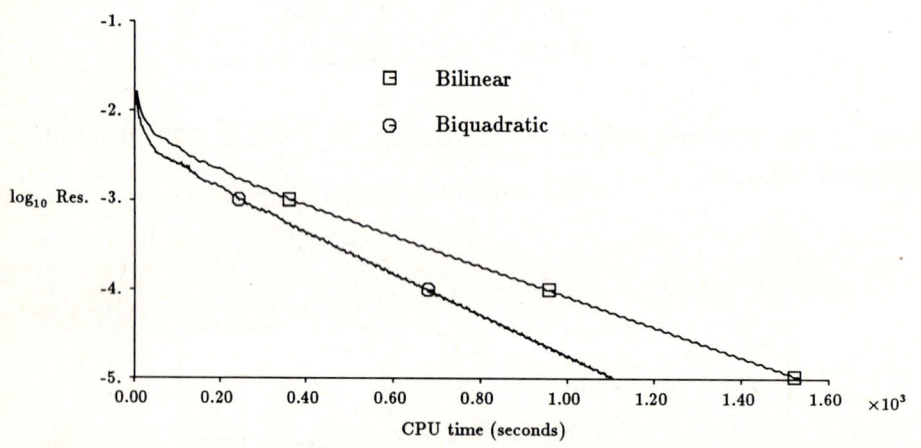

Figure 5.45: CPU Comparison, 10% $\cos^2$ Bump, $M_\infty = 0.5$, Biquadratic and Bilinear Elements

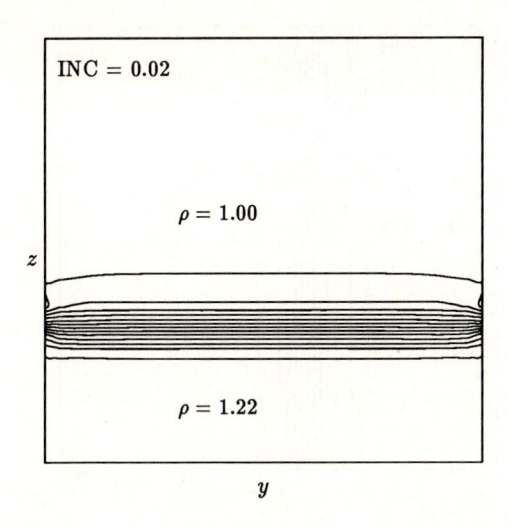

Figure 5.46: Density, 5° Wedge, $M_\infty = 2$, Y-Z Slice at X = 0.5

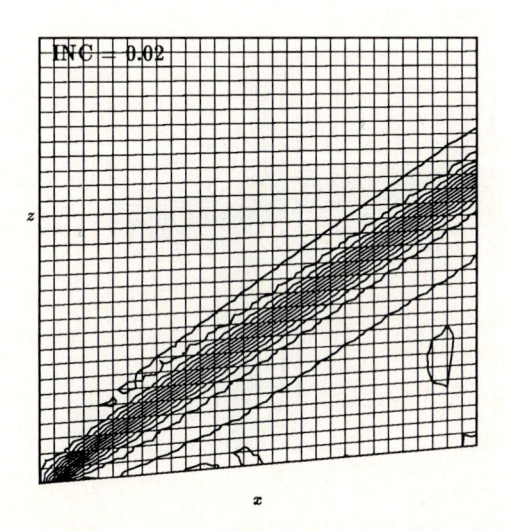

Figure 5.47: Pressure, 5° Wedge, $M_\infty = 2$, X-Z Slice at Y = 0.5

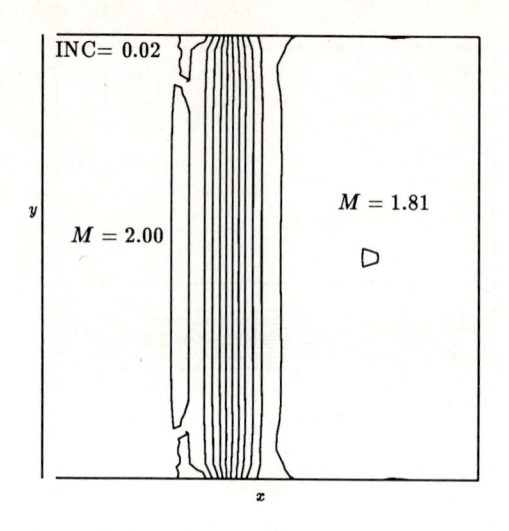

Figure 5.48: Mach Number, 5° Wedge, $M_\infty = 2$, X-Y Slice at $Z = 0.7$

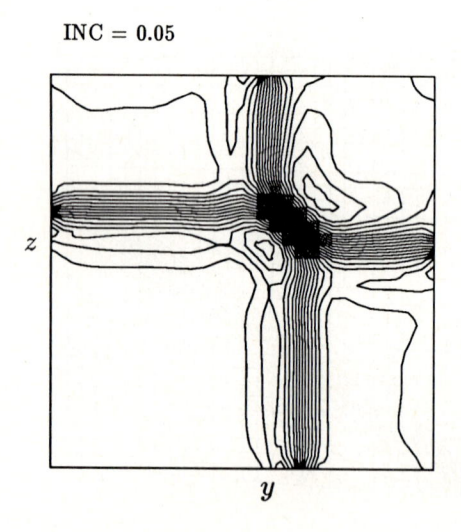

Figure 5.49: Density, 10° Double Wedge, $M_\infty = 2.5$, Y-Z Slice at $X = 0.67$

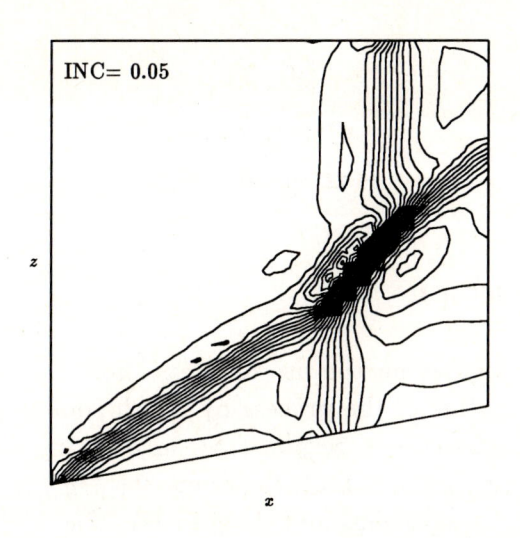

Figure 5.50: Mach Number, 10° Double Wedge, $M_\infty = 2.5$, X-Z Slice at Y = 0.5

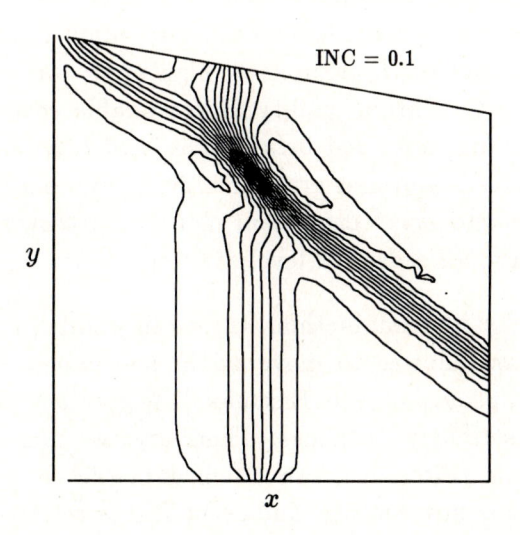

Figure 5.51: Pressure, 10° Double Wedge, $M_\infty = 2.5$, X-Y Slice at Z = 0.7

Chapter 6
Adaptation

6.1 Introduction

This chapter focuses on the use of adaptation in solving the Euler equations. Adaptation is the process by which some aspect of the solution algorithm changes in response to an evolving solution. These changes can be in the governing equations [11], in the computational grid, or in both the equations and the grid [29]. In this thesis, adaptation refers to changes in the computational grid as the solution proceeds. In a problem, certain regions of a computational domain will have "interesting" features (shocks, boundary layers, recirculation zones, etc.), while other regions will have smooth and relatively uninteresting flow. The interesting regions are often regions with high gradients and hence larger numerical errors. The idea behind grid adaptation is to increase the number of points in the regions of high error (or gradient) to try and increase overall solution quality at reasonable cost. Several ways of adapting grids are described below. The *grid redistribution* method attempts to get to an optimal grid for a given number of points. The *grid regeneration* and *grid enrichment* methods attempt to reduce the computational cost for a desired level of error.

In the grid redistribution method, the mesh points in an initial computational grid are allowed to move as the solution proceeds. Survey articles by Thompson [70] and Eiseman [13] give a good overview of various grid redistribution schemes. There are two main advantages of grid redistribution. First, one can often align the computational grid with the interesting flow feature, and second, it is relatively easy to implement these methods in an existing structured (or unstructured) grid code. The main disadvantages are that elements tend to become highly distorted, and that it is often difficult to know *a priori* how many points are needed in an initial grid.

A second approach sometimes taken is grid regeneration. In this method, either the entire computational grid or some smaller portion of it is regenerated when adaptation occurs. The grid point distribution is determined by some error indicator based on the solution on the previous grid. This method is described in detail in articles by Peraire, et al. [53, 54]. This retains the grid alignment advantages while reducing grid skewness problems. The main disadvantages are that the cost of grid regeneration is usually fairly high, the interpolation of the solution from the old grid to the new grid can be difficult, and the method (so far) has been limited to triangular and tetrahedral elements.

The third approach is grid enrichment. The survey article by Berger [4] presents a good overview of grid enrichment methods. In this approach, additional nodes and elements are inserted into the grid. There are two main subclasses of these embedded mesh methods: those in which the embedded mesh is not aligned with the initial mesh [5] and those in which the mesh is aligned with the initial mesh [20, 40, 72, 73]. Under the latter subclass of embedded mesh methods, one can further distinguish between methods that result in interface regions [21, 52, 63] and those that have no interfaces [48, 58]. In this thesis, the grids are aligned with the initial mesh and have interfaces which require special treatment.

The next section of this chapter will present an explanation of the procedure used to determine where and how to adapt. Then the interface treatment will be discussed, and some properties of these interfaces examined. Some examples demonstrating the utility of adaptation will be presented, and the computational savings produced by adaptation will be examined.

6.2 Adaptation Procedure

In a typical problem, one starts out with an idea of what the flow will look like, but one usually does not know exactly where the features lie. The adaptive approach starts with an initial grid coarse enough to be inexpensive to compute, yet fine enough so that most of the essential features can appear. The first step in the adaptive solution procedure is to compute a solution on this coarse grid. It is usually not necessary to allow this solution to converge completely, and in the method describe here the solution is allowed to evolve "halfway" to convergence, that is, until the convergence parameter reaches the square root of its converged value. Then, an adaptation parameter is calculated, and cells are flagged either for refinement or unrefinement. After the initial flagging, elements adjacent to the cells marked for division are marked for division. This flagging of adjacent cells can be continued for any number of passes to enlarge the adapted region. In the test cases computed, only the adjacent cells need to be flagged to obtain good results. The flagged cells at the coarsest level are divided, then those at the next coarsest, and so on. After all division is done, unrefinement proceeds from the finest level to the coarsest. After the initial set of refinements and unrefinements in performed, the mesh is scanned for "holes". In two dimensions, an element is considered part of a hole if 3 or more of its faces have been subdivided. In three dimensions, an element is considered a hole if 4 or more of its faces are subdivided. If an element is a hole, then it is subdivided to eliminate the hole. This can result in the formation of more holes, so the process is repeated until the grid stops changing. The analogous "island" of fine cells in the midst of a region of coarse cells is not treated explicitly as the adjacent element flagging algorithm prevents the formation of excessively small islands in most cases. In any case, islands are less important than holes, since a hole in a fine region may result in reduced accuracy, but an island in a coarse region should only increase or not change the accuracy of the algorithm.

After all mesh changes have been made, the grid geometry is recalculated, along with some quantities used for vectorization. The solution is interpolated onto the new grid, and the calculation proceeds. In a typical problem, an adaptation take as much time as several iterations, but since adaptation occurs infrequently, the time involved is not significant.

Figure 6.1 shows an example of how elements are refined and unrefined in two dimensions. The elements marked with the • are each subdivided into 4 elements. The elements marked with the ○ are fused back to form one element. This figure also illustrates the main rule about which elements can be refined or unrefined. This rule is simply that the refining or unrefining of an element cannot result in more than one interface node on the face of any element. For example, the element marked with the ×

Before　　　　　　　　　　　After

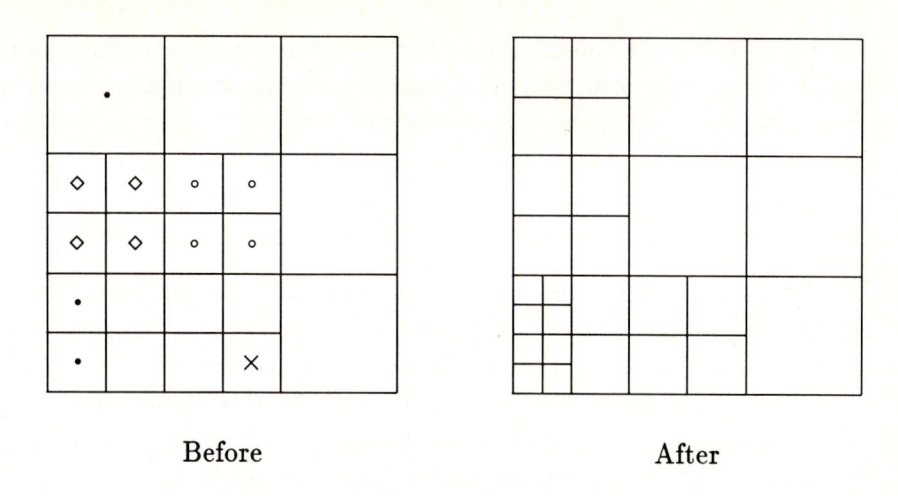

Figure 6.1: Example of Mesh Adaptation

cannot be refined, since this would result in two additional nodes on the face of the element to its right. The elements marked with the ◇ cannot be unrefined, since the lower face would then have too many nodes on it. In three dimensions, the comparable rules are that no edge can have more than one additional node on it, and each face can have at most one additional mid-face node. A discussion of some of details of the data structure used to perform cell division and restoration may be found in Section A.4.

6.2.1 Placement of Boundary Nodes

During adaptation, new nodes may be created on boundary surfaces. One of the problems with this is that one would like to increase the resolution of the body, as well as the resolution in the interior. If one picks the boundary node coordinates to be the midpoint of the edge being divided, one can easily obtain a grid which calculates the flow over a "faceted" body, rather than a smooth body. The approach taken to solve this problem is to fit a cubic which passes through the two nodes on the face being divided, with the normal to the curve at those points defined by the normal vector at the nodes. The new mid-face node is placed on this curve. For certain geometries, this cubic fitting is undesirable (if all surfaces are straight lines, for example), so it is possible to disable cubic fitting and use simple linear fitting. In three dimensions, only linear fitting is used, which is sufficient for the geometries treated in this thesis.

6.2.2 How Much Adaptation?

The decision on how many levels to use and how many times to adapt is completely in the hands of the user. In principle, the algorithm allows any number of adaptation levels, but for most problems examined in this thesis either 3 or 4 levels is sufficient. In this thesis, the main factor determining the number of levels used in 2-D was the resolution of the printer, and in 3-D the amount of memory available on the computer. For practical computations, a more physical criterion such as convergence of lift [20], might be more useful.

How many times to adapt is a different issue. For steady problems, the number of elements tends to increase with succeding adaptations until a maximum is reached, usually at the adaptation in which the finest grid level is first introduced. Succeeding adaptations tend to remove excess elements (especially near shocks), until some "converged" grid is reached. The grid plots in section 6.5.1 illustrate this convergence of the adapted grids. Usually, adapting twice after the final grid level is introduced is sufficient to obtain this convergence.

6.3 Adaptation Criteria

In order to decide which cells to divide or undivide, one needs to define some sort of adaptation parameter. There is a good amount of literature indicating possible choices for this adaptation parameter. Dannenhoffer and Baron [21] suggest using the first difference of density as the criterion for adaptation. Löhner [39] suggests using a more complex indicator, which is essentially a second difference of density scaled by a first difference of density. Berger, et al. [5] use Richardson extrapolation to estimate the truncation error in the solution, and refine to evenly distribute the error. The test cases presented and others not presented here indicate that the first difference of density tends to give better grids for transonic flows, while the scaled second difference works better for supersonic flows. The Richardson extrapolation type indicators were not implemented in the present study.

6.3.1 First-Difference Indicator

The first difference indicator is empirically derived [20] and based on the undivided first difference of density. It is calculated as follows:

1. In each element, calculate the absolute value of the first difference of density. For bilinear elements, the quantity used is

$$A_e = \max(|\rho_2 - \rho_1|, |\rho_3 - \rho_2|, |\rho_4 - \rho_3|, |\rho_1 - \rho_4|) , \qquad (6.1)$$

    where ρ_i is the density at element node i. For trilinear elements, a similar quantity is defined.

2. Compute the mean and standard deviation of this quantity.

3. Normalize this quantity by subtracting the mean and dividing by the standard deviation. If the standard deviation is small, use some arbitrary value instead (currently 0.05 times the mean or .0005, whichever is larger.)

4. Compute the median, and subtract it from the parameter.

5. If the scaled parameter is greater than some threshold value, try to refine that element. If it less than another threshold value, try to unrefine the element.

Refine thresholds near .3 and unrefine thresholds near $-.3$ give good results for many problems. The shifting by the median is performed after the normalization to make the calculation of the median easier, and the shifting is necessary because the distribution of the adaptation parameter can become very skewed as the calculation progresses. In some problems, this skewness results in over-coarsening of the mesh at later adaptation cycles.

6.3.2 Second-Difference Indicator

The second-difference based parameter is computed as follows:

1. Compute a nodal second difference of density using one of the methods outlined in Section 4.5 above (usually the high-accuracy method).

2. Compute a nodal first difference. For both bilinear and trilinear elements, the following is used:

$$\text{First Difference}_i = \frac{1}{n_i} \sum_j |\rho_i - \rho_j|, \qquad (6.2)$$

where j ranges over all nodes in all elements containing node i, and n_i is the number of elements containing node i.

3. Compute an average density at the nodes.

4. Compute a nodal adaptation switch:

$$A_n = \frac{|\text{Second Difference}|}{|\text{First Difference}| + \epsilon \times \text{Average}}, \qquad (6.3)$$

where ϵ is a small parameter (of order 0.05) added to "smooth" the switch in smooth portions of the flow. Without it, small oscillations in the flow tend to produce large values of the switch.

5. Distribute the nodal switch to the elements.

6. If the switch is greater than some value (usually around 0.15), divide; if it is less than some value (usually around 0.05), undivide.

This indicator is based on results from interpolation theory which indicate that the interpolation error in a linear scheme is proportional to the second derivative of the quantity being interpolated. The first difference is used as a "fudge factor" to reduce the sensitivity of the switch to greatly differing shock strengths. Without the first difference scaling, the stronger shocks in the flow tend to receive most of the adaptation, while the weaker ones receive very little. Other advantages of this indicator are explained by Löner in [39].

For this indicator, an automatic threshold method was tried. This sets the unrefine threshold to be some fraction of the RMS value of the adaptation indicator, and sets the refine threshold to be some multiple of the RMS value. Experimentally, values of the fraction and multiple near 0.2 and 1.1 seem to work well.

6.3.3 Two-Dimensional Directional Adaptation

In many cases, the features of interest have strong variations in only one direction. In these cases, it is wasteful to introduce points in the direction along which little variation occurs. This leads to the concept of directional adaptation. The original motivation in [29] was to use this idea for viscous flows with boundary layers. The extension of these ideas to inviscid flow was attempted. A significant difference between the viscous and inviscid flows is that in the viscous flows features like boundary layers tend to be aligned with the grid, while in the inviscid flows, the features (mainly shocks) lie in all directions. A criterion for determining which way to divide a cell was developed. The first step is to use some criterion for determining which cells to subdivide. After this is done, determine whether to divide into four cells or two cells with the following procedure. Calculate

$$\Delta_H = |\rho_4 - \rho_1| + |\rho_3 - \rho_2|, \tag{6.4}$$

$$\Delta_V = |\rho_2 - \rho_1| + |\rho_3 - \rho_4|, \tag{6.5}$$

$$D_H = \frac{\Delta_H}{\Delta_H + \Delta_V}, \tag{6.6}$$

$$D_V = \frac{\Delta_V}{\Delta_H + \Delta_V}. \tag{6.7}$$

If D_H is greater than some threshold (typically .2–.4), divide the cell horizontally (parallel to the 1-2 face). If D_V is greater then some threshold, divide the cell vertically (parallel to the 2-3 face). If both exceed their thresholds, or neither exceeds its threshold, divide the cell into 4 subcells as for regular embedding.

In practice, the directional embedding has not proven to be very useful for the Euler equation test cases examined [63]. In order to use directional embedding to the greatest advantage, the flow features must be aligned with the grid. In inviscid flow, this is often not the case. In Navier-Stokes calculations, however, directional embedding has been applied successfully by Kallinderis and Baron [29]. A combination of the grid redistribution and grid embedding ideas may make directional embedding much more useful.

6.4 Embedded Interface Treatment

The use of quadrilaterals in two dimensions and hexahedra in three dimensions results in "hanging" nodes at the interfaces between the coarse and fine regions of the mesh. For the remainder of this discussion, these nodes will be called *virtual* nodes. These virtual nodes must be properly treated to obtain a stable, consistent and conservative scheme. It is possible to avoid interfaces altogether by mixing triangular and quadrilateral elements [58], but that approach is not used here.

6.4.1 Two-Dimensional Interface

In two dimensions, an element face may have at most one additional node on it. Figure 6.2 shows a typical interface between a locally fine region and a coarser region. The fluxes at the interface node (node 2) are set to the average of the fluxes at nodes 1 and 3 before residuals are calculated. The state vector at node 2 is set to the average of the state vectors at nodes 1 and 3 after each iteration. No other special treatment is applied at nodes 1,2, or 3. Thus, the overhead involved in having interfaces is very small, amounting to less than 2% of the total time per iteration for a typical mesh. This step is also fully vectorized, so it is not a limitation on machines with large vector/scalar speed ratios.

Conservation can be important in adaptive meshing, since one may have shocks passing through or near interfaces. Repeating the argument of section 4.7, for an interior, non-virtual node j, each entry in R_x is of the form

$$\int \tilde{\mathbf{N}}_i \frac{\partial \mathbf{N}_j}{\partial x} dV, \qquad (6.8)$$

so the column sum (which represents the total contribution of node j)

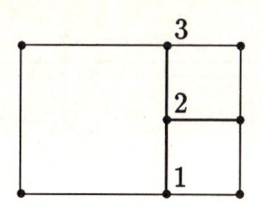

Figure 6.2: Detail of Two-Dimensional Interface

is

$$\int \sum_i \tilde{\mathbf{N}}_i \frac{\partial \mathbf{N}_j}{\partial x} dV. \tag{6.9}$$

Integrate Eq. (6.9) once by parts to obtain

$$\int \sum_i \tilde{\mathbf{N}}_i \frac{\partial \mathbf{N}_j}{\partial x} dV = \oint_{\partial V} \sum_i \tilde{\mathbf{N}}_i \mathbf{N}_j \hat{n} \cdot dS - \int \mathbf{N}_j \sum_i \frac{\partial \tilde{\mathbf{N}}_i}{\partial x} dV. \tag{6.10}$$

The second term on the right-hand side is zero, since the test functions are chosen so that their sum is 1. The line integral in Eq. (6.9) can be written as the sum of elemental line integrals, as shown in Fig. 6.3. Since $\sum \tilde{\mathbf{N}}_i = 1$ in each element, the line integral simplifies to

$$\sum_e \oint_{\partial V_{(e)}} N_j^{(e)} \hat{n} \cdot dS. \tag{6.11}$$

On the boundary of the region, $N_j = 0$, so if all the interior contributions cancel, the complete integral will be zero. All that is required to get cancellation is that the interpolation functions in each element are continuous across the element faces. This is clearly true on all faces except the line jB, and it will be shown that it is true along this line as well. The trick is to make a proper interpretation of N_j in elements 1 and 2.

Since the flux at node A is the average of the fluxes at nodes j and B, one can eliminate node A by adding half of its *interpolation* function to each of nodes B and j, so that in elements 1 and 2, a new N_j', valid in elements 1 and 2, is defined to be

$$N_j' = N_j + \frac{N_A}{2}, \tag{6.12}$$

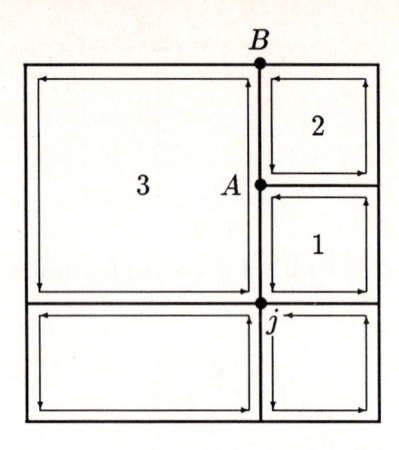

Figure 6.3: Line Integrals for Conservation at Interfaces

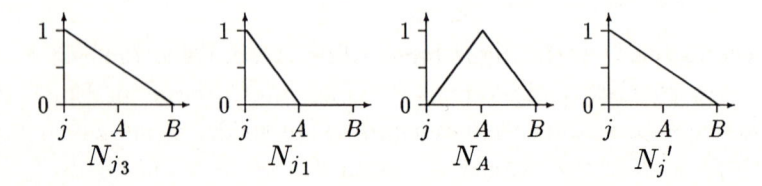

Figure 6.4: Interpolation Function N_j in Elements 1,2 and 3

where N_j is the standard interpolation function, non-zero only in element 1. Now it is only necessary to show that $N_{j3} = N_j'$ along the line jB. Figure 6.4 shows N_{j3} and N_j' indicating that this equivalence holds, so the total contribution of any node to the flux residuals is zero, as required for conservation. This is not the full story, however. Since after each iteration, the changes at node A are discarded, strict conservation is no longer maintained. One can remedy this by distributing the residual at node A to nodes j and B, but for steady problems, the effect of doing this redistribution is slight. Typical conservation errors introduced by not performing this distribution range from 0.1% to 0.5% for the problems in this thesis.

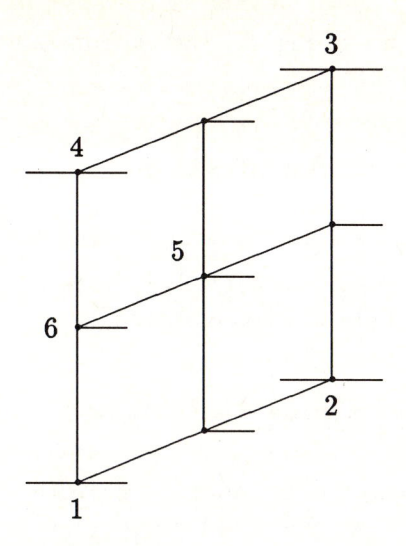

Figure 6.5: Cutaway View of Three-Dimensional Interface

6.4.2 Three-Dimensional Interface

In three dimensions there are two types of interface nodes: mid-face nodes and mid-edge nodes. Figure 6.5 shows a cutaway view of the interface between a fine and coarse region in three dimensions. In this figure, the fine region is to the right, and the coarse region is to the left. The fluxes and state vectors at a mid-edge node (node 6) are the averages of the fluxes and state vectors at the ends of the edge (nodes 1 and 4 in the figure). The fluxes and state vectors at the mid-face node (node 5) are the averages of the fluxes and state vectors at the corner nodes (nodes 1-4). This procedure is fully vectorizable, and the overhead in the interfaces is again negligible.

One consequence of the interface treatments used in 2- and 3-D is that certain plot quantities can be discontinuous across the interface. For example, Mach number and pressure are non-linear combinations of the continuous state vectors, so it is possible for discontinuities to occur. This discontinuity does not appear in the density contours, since density is a linear combination of the state variables. Figure 6.6 shows Mach number contours illustrating this kind of discontinuity. Note that in the density contours of Fig. 6.7 (taken from the same set of data) no discontinuity is present.

6.5 Examples of Adaptation

This section presents some examples of the use of adaptation. The test problems of the previous chapter will be recomputed using adaptive gridding, and the results will be compared.

6.5.1 Multiple Shock Reflections

To verify the accuracy of the adaptive algorithm, the shock reflection cases of Section 5.2 were computed using adaptive gridding. For all of these test cases the cell-vertex method was used.

The 5 degree, $M_\infty = 2$ case was computed using the second difference switch and the automatic thresholding described above, with the unrefine threshold set to 0.25 and the refine threshold set to 1.15. The initial grid was 32x8 and the grid was adapted five times. Only three levels of cells were permitted so on the third, fourth and fifth adaptations no new levels of cell were added. The finest level corresponds to a 128x32 global grid. Figs. 6.8-6.13 show the base grid and the grids after each of the adaptations steps. Note how the finest region confines itself more and more to the region around the shocks as the adaptation proceeds. Also note that the grid after the fifth adaptation is very similar to the grid after the fourth adaptation, indicating that the adaptive procedure has "converged." The final mesh has about 1600 elements, compared to the 4100 elements of a globally fine grid, for a savings of a factor of 2.6. The savings is low mainly because so much of the domain is in the vicinity of a shock. The scramjet calculations of chapter 8 also illustrate this behavior. As the number of adaptation levels increases, the savings become larger for this type of problem, since the adaptation is confined to a smaller region of the flow. Figure 6.14 shows the Mach number contours for this case, with interfaces between coarse and fine regions indicated by the dotted lines. Note that the shocks are sharper than the shocks in Fig. 5.6 even though both cases used approximately the same number of elements. Here, sharper means that the feature is spread out over a smaller region in physical space, even if it take up the same number of points in computational space. Also note that the fine region just surrounds the shocks, indicating that the adaptation indicator is correctly responding to the feature. Figure 6.15 shows the density on the line $y = 0.6$, along with the exact solution. Note the accuracy of the solution and the sharper shocks. This case required 203 seconds to compute on the Alliant, compared to 375 seconds for a globally fine (128x32) case.

The 15 degree, $M_\infty = 4$ test problem was also computed. Three grid levels were allowed, and the grid was adapted four times, using the second difference adaptation indicator. Figure 6.16 shows the final grid for this problem. Notice that the grid surrounds the shock completely. Figure 6.17 shows pressure contours for this problem. The shock is quite clean and the reflection is captured accurately. Figure 6.18 shows the density on a slice at $y = 0.9$ along with the exact solution. The agreement with the globally fine solution in Fig. 5.21 is very good. This case (with the finest grid the equivalent of a 88x32 grid) required 84 seconds to compute, compared to 104 seconds for the the 66x24 global grid.

6.5.2 4% Circular Arc Bump

The 4% circular arc bump with $M_\infty = 1.4$ was computed using both indicators discussed above. The final grids for these cases are shown in Fig. 6.19 (first difference indicator) and Fig. 6.20 (second difference indicator). These grids have 1698 and 1620 elements. Figures 6.21 and 6.22 show the pressure contours for these two cases. Aside from a small difference in the number of elements, the solutions look approximately the same, with the reflected shocks in the same positions, etc. The real difference between the two indicators is only apparent when one looks at the computational time required to obtain the solution. The second difference indicator case required 448 iterations and 215 CPU seconds to converge, while the first difference indicator required 640 iterations and 332 second to reach convergence. This discrepancy arises at the intermediate grid adaptation stages, where the first difference indicator produces grids which do not surround the shocks adequately. Figure 6.23

shown the convergence histories for the two cases. Note that in the first difference case, there is a long period of little convergence, which corresponds to the shocks re-adjusting themselves to a new grid topology. The second difference indicator captures the shocks immediately, and so converges much more rapidly. This case illustrates the slight superiority of the second difference indicator over the first difference indicator for supersonic flows. The second difference indicator is a much sharper shock detector, and is also less sensitive to shock strength. These two features suggest the use of the second difference indicator for supersonic flows, while the next example shows that for transonic flows the first difference indicator is preferable.

6.5.3   10% Circular Arc Bump

The 10% circular arc bump with $M_\infty = 0.68$ was computed using the two indicators discussed above. The final grids for these cases are shown in Fig. 6.24 (first difference indicator) and Fig. 6.25 (second difference indicator). These grids have 990 and 1011 elements. Figures 6.26 and 6.27 show pressure contours for the first difference and second difference indicators. The solutions obtained are both good (compare with Fig. 5.27) and show the shock in the right place, with relatively smooth contours over the bump. In this case, the first difference indicator required 469 iterations and 128 seconds to converge, compared with the 517 iterations and 166 seconds required by the second difference indicator. In this case, the time discrepancy is due to the fact that the intermediate grids in the second difference case had more points than the first difference indicator grids, resulting in more intermediate computation. Since the shock is captured immediately by both indicators, these additional elements did not contribute to more rapid convergence as they did for the supersonic bump above.

6.5.4   3-D Channel

To check the three-dimensional adaptation, the $M_\infty = 2.5$, 10° double wedge of the previous chapter was computed using 3 grid levels. The initial grid was 12x12x12 and was adapted 4 times, with the final grid containing 57714 elements. A globally fine grid would contain about

110,592 elements, for a savings of a factor of 2. This small factor is due to the fact that most of the domain is near a shock, resulting in adaptation nearly everywhere. Figure 6.28 shows the density on the slice $x = 0.67$. Note the curvature of the shock surfaces as they intersect. Figure 6.29 shows the Mach number on a slice at $y = 0.5$ and Fig. 6.30 shows pressure on a slice at $z = 0.7$. These figures illustrate all the essential features of the flow, and demonstrate the soundness of the algorithm. Figure 6.30 also shows a slice of the grid. Notice the clustering of the fine cells around the shock surfaces. Unfortunately, there is a lack of adaptation at the intersection of the shocks, indicating that the adaptation criterion needs further investigation. This case required just under 3 hours on a single-processor IBM 3090 with a vector facility.

6.5.5 Distorted Grid

Adaptation can also allow one to obtain good results with poor initial grids. Fig. 6.31 shows a distorted grid for the 10% bump case. The finite element solution was computed using exactly the same input parameters as for the first difference indicator test case above. Figure 6.32 shows the final grid containing 1095 elements and Fig. 6.33 shows pressure contours. The solution compares well with the solutions obtained on non-distorted grids, illustrating the robustness of the adaptive algorithm. This example completes the verification of the adaptive algorithm, which will be put to use in the scramjet solutions of chapter 8.

6.6 CPU Time Comparisons

This section presents a short comparison of the computational work for various solution approaches (adaptive, biquadratic, etc.) for the $M_\infty = 1.4$, 4% bump, the $M_\infty = 0.68$, 10% bump, and the $M_\infty = 5$ two-dimensional scramjet (section 8.2).

Table 6.1 shows the number of elements, number of iterations and total Alliant CPU time for the $M_\infty = 1.4$, 4% bump case. The adapted bilinear case was computed using the second difference switch described above. The most interesting thing about this set of data is that the biquadratic elements require the least computational effort, saving about

Table 6.1: CPU Comparisons for $M_\infty = 1.4$, 4% Bump

Case	# Elements	Iterations	CPU Seconds
120x40 Bilinear	4800	386	592
Adapted Bilinear	1620	448	215
30x10 Biquadratic	300	331	131

Table 6.2: CPU Comparisons for $M_\infty = 5$ scramjet

Case	# Elements	Iterations	CPU Seconds
Globally Fine (3)	26624	400 (est.)	3000 (est.)
Adapted Bilinear	12998	722	2250
Globally Fine (2)	6656	257	482
Biquadratic	416	140	76

a factor of 4.5 over the global bilinear mesh. Also of note is the fact that the ratio of adapted CPU time to globally fine CPU time is only about 1:3. In flows dominated by shocks, there tends to be so much adaptation that the savings are not that dramatic. The scramjet example following also illustrates this.

Table 6.2 gives the CPU statistics for the $M_\infty = 5$ scramjet example. There are two globally fine times reported in the table—the 6656 element case represents a two-level global refinement of the 416 element initial grid used for the adaptive calculations, and the 26624 element times are estimates for a three-level refinement based on scaling the number of iterations and the time per iteration. This very fine grid corresponds to the finest level in the adapted case. Note again that the biquadratic elements emerge as the fastest solution technique. The accuracy from the biquadratic elements is comparable to the 6656 element accuracy, but there is a cost savings of a factor of 6. Again, note that adaptation does not save as much effort, with a factor of 2 savings in storage and an estimated CPU savings of only 1.3.

The place where adaptation saves the most time is in transonic calculations. Table 6.3 shows the comparison between three test cases used

Table 6.3: CPU Comparisons for $M_\infty = 0.68$, 10% Bump

Case	# Elements	Iterations	CPU Seconds
120x40 Bilinear	4800	1391	2097
30x10 Biquadratic	300	875	343
Adapted Bilinear	990	469	128

to compute the transonic bump problem. In this case, adaptation saves a factor of 16 in CPU time over the globally fine mesh. These results are comparable to the results reported by Dannenhoffer [20]. Surprisingly, the biquadratic elements do not fare as well compared to the adapted case as they do above, but the biquadratic elements still provide a CPU savings of a factor of 6 over the globally fine grid. The primary conclusion from these comparisons is that adaptation can produce either very large or modest saving, depending on the details of the problem. For problems with a great deal of structure (such as the scramjet), adaptation saves the least. For transonic problems, however adaptation is extremely useful.

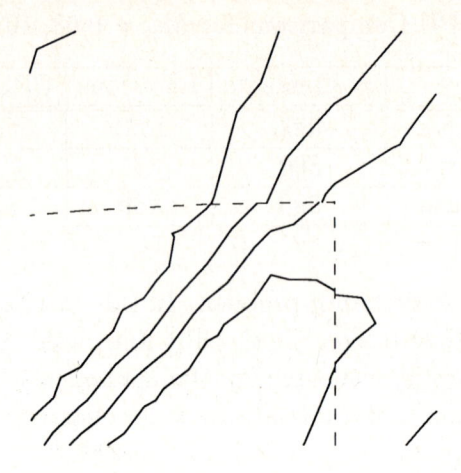

Figure 6.6: Mach Number Contours Showing Interface Discontinuity

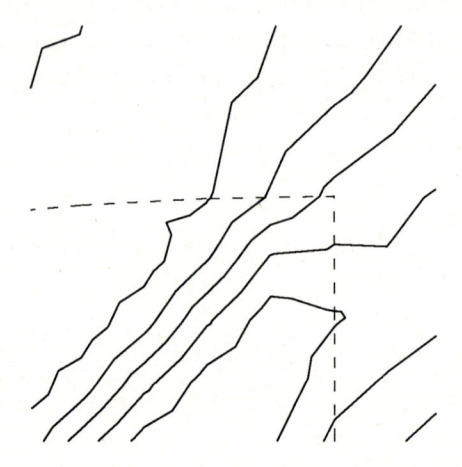

Figure 6.7: Density Contours Showing No Interface Discontinuity

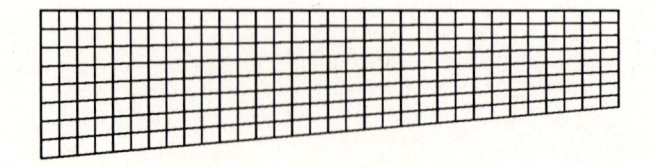

Figure 6.8: Base Grid, 5° Wedge, 256 Elements, 297 Nodes

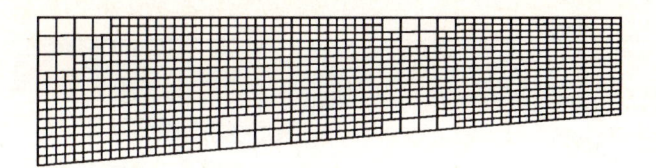

Figure 6.9: Grid After 1 Adaptation, 937 Elements, 1024 Nodes

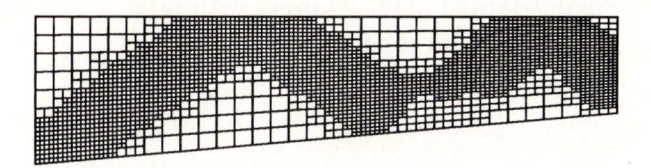

Figure 6.10: Grid After 2 Adaptations, 2551 Elements, 2748 Nodes

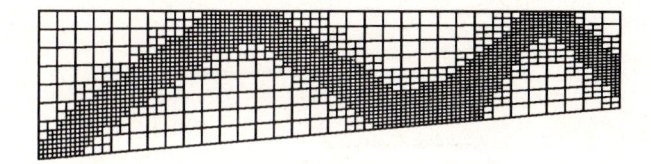

Figure 6.11: Grid After 3 Adaptations, 1798 Elements, 2005 Nodes

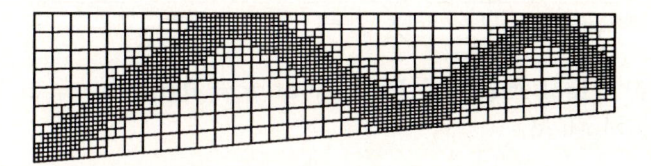

Figure 6.12: Grid After 4 Adaptations, 1600 Elements, 1808 Nodes

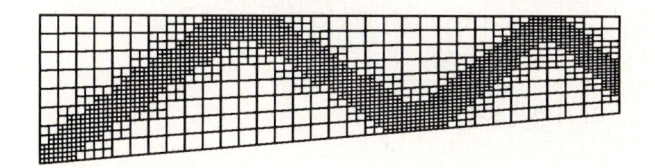

Figure 6.13: Final Grid After 5 Adaptations, 1591 Elements, 1799 Nodes

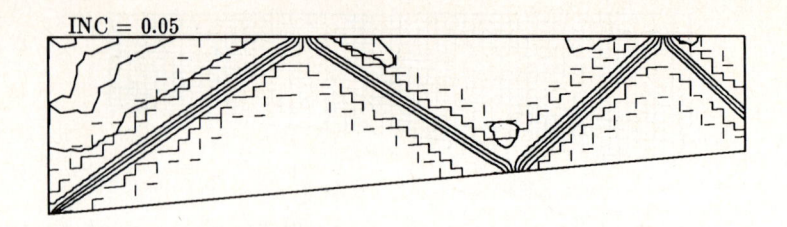

Figure 6.14: Final Adapted Mach Number with Interfaces Shown, $M_\infty = 2$, 5° Wedge, Cell-Vertex Method

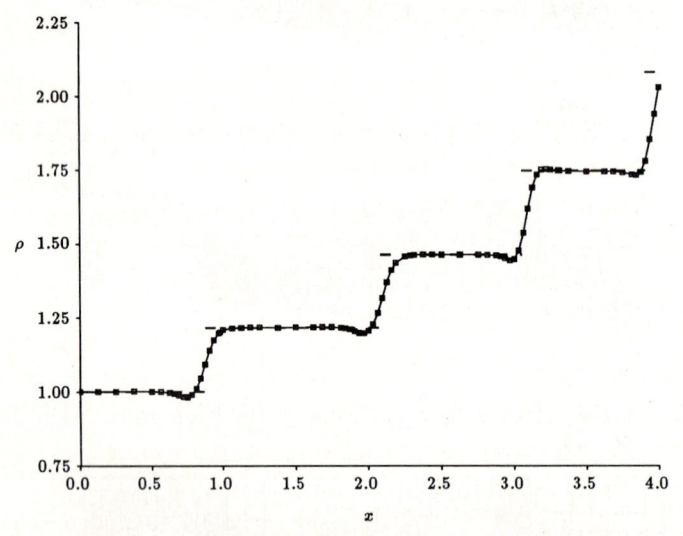

Figure 6.15: Final Adapted Density at $y = 0.6$, $M_\infty = 2$, 5° Wedge, Cell-Vertex Method

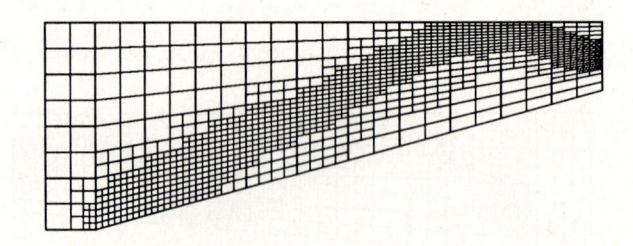

Figure 6.16: Final Grid After 4 Adaptations, 15° Wedge, 1148 Elements, 1274 Nodes

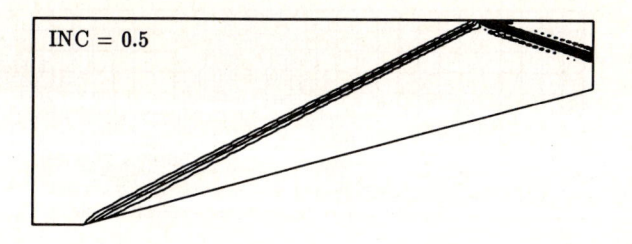

Figure 6.17: Pressure, $M_\infty = 4$, 15° Wedge, 4 Adaptations

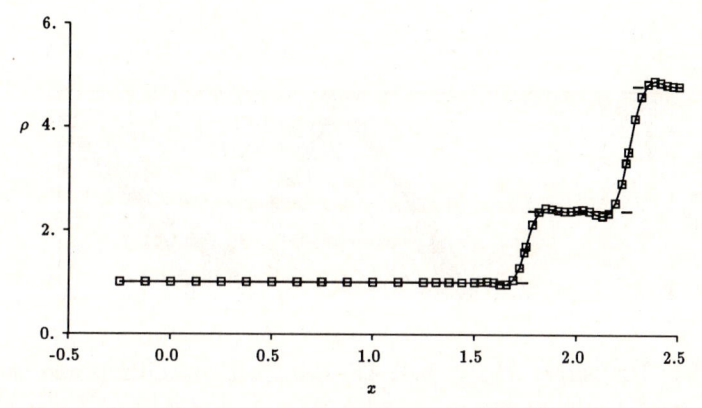

Figure 6.18: Density at $y = 0.9$, $M_\infty = 4$, 15° Wedge, 4 Adaptations

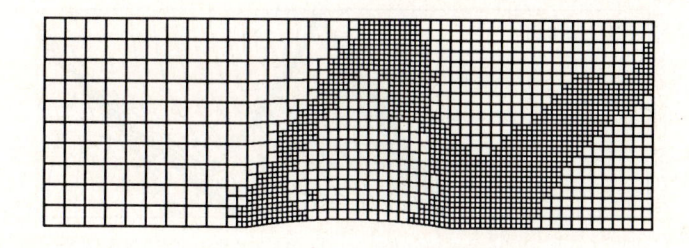

Figure 6.19: Final Grid, $M_\infty = 1.4$, 4% Bump, First Difference Indicator, 1698 Elements

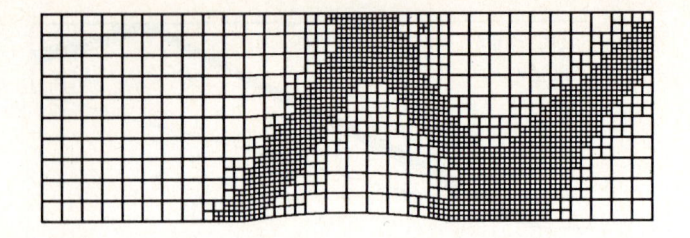

Figure 6.20: Final Grid, $M_\infty = 1.4$, 4% Bump, Second Difference Indicator, 1620 Elements

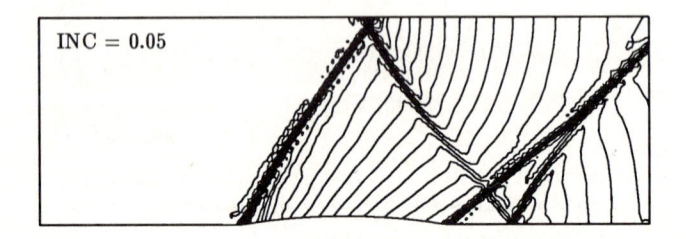

Figure 6.21: Pressure, $M_\infty = 1.4$, 4% Bump, First Difference Indicator

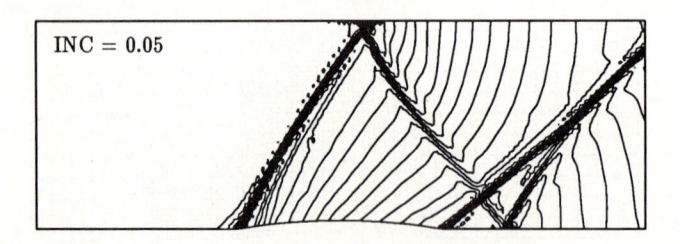

Figure 6.22: Pressure, $M_\infty = 1.4$, 4% Bump, Second Difference Indicator

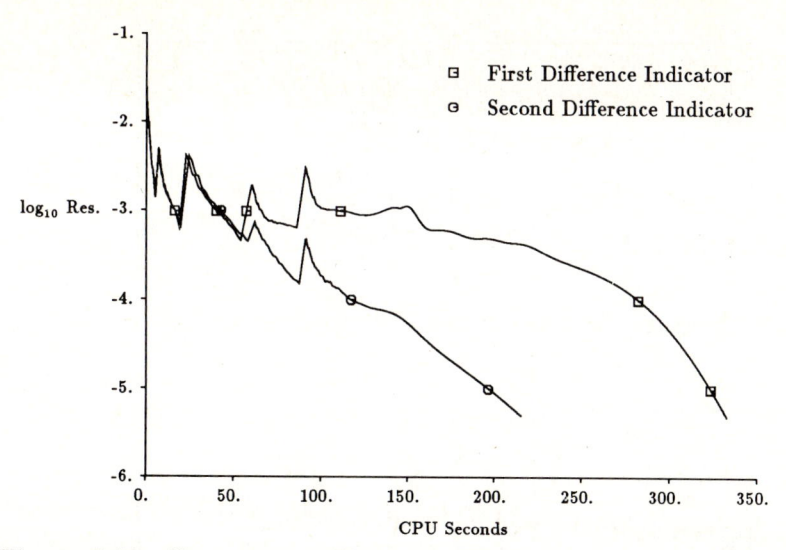

Figure 6.23: Convergence Histories for Both Indicators, $M_\infty = 1.4$, 4% Bump

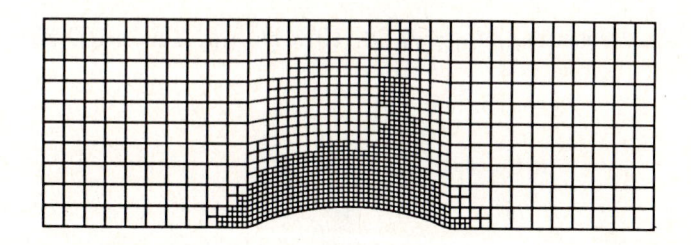

Figure 6.24: Final Grid, $M_\infty = 0.68$, 10% Bump, First Difference Indicator, 990 Elements

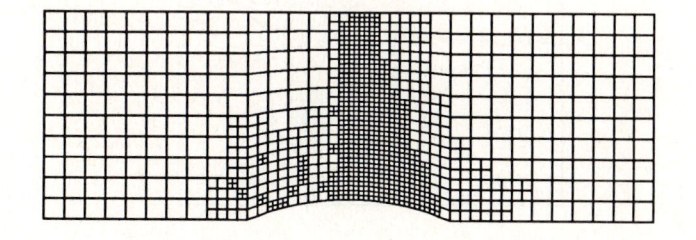

Figure 6.25: Final Grid, $M_\infty = 0.68$, 10% Bump, Second Difference Indicator, 1011 Elements

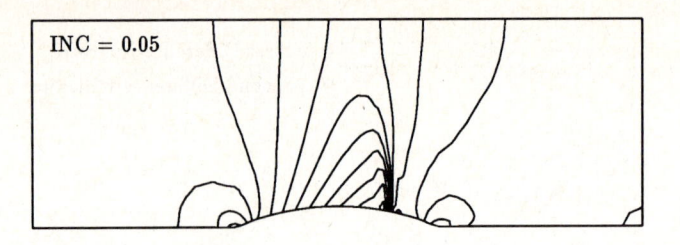

Figure 6.26: Pressure, $M_\infty = 0.68$, 10% Bump, First Difference Indicator

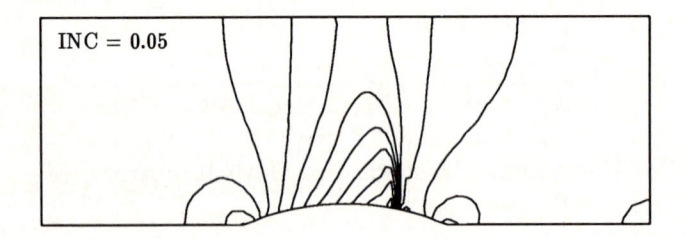

Figure 6.27: Pressure, $M_\infty = 0.68$, 10% Bump, Second Difference Indicator

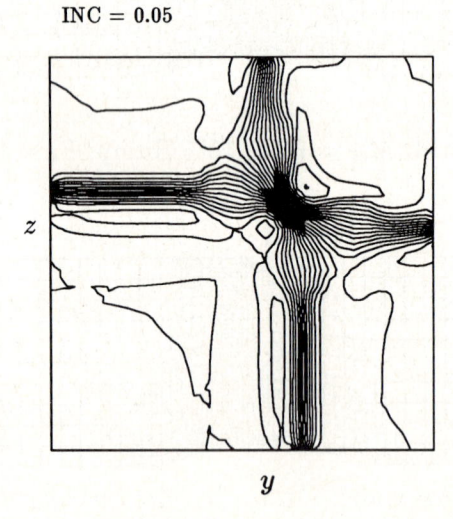

Figure 6.28: Density at $x = .67$, $M_\infty = 2.5$, 10° Double Wedge, 57714 Elements

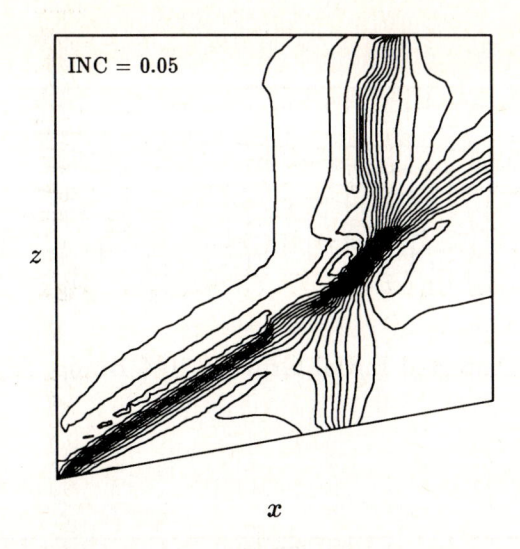

Figure 6.29: Mach Number $y = .5$, $M_\infty = 2.5$, 10° Double Wedge, 57714 Elements

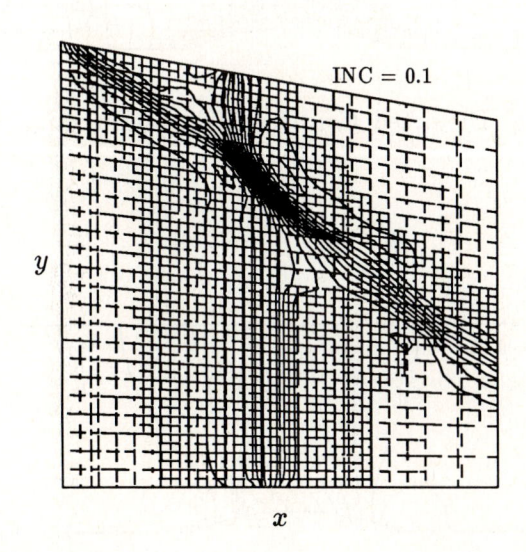

Figure 6.30: Pressure at $z = .7$, $M_\infty = 2.5$, 10° Double Wedge, 57714 Elements (Grid Dotted)

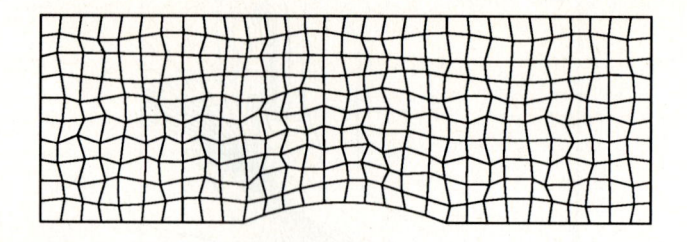

Figure 6.31: Distorted Initial Grid for 10% Bump, $M_\infty = 0.68$

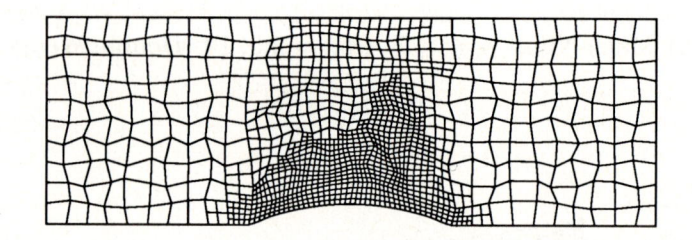

Figure 6.32: Final Distorted Grid for 10% Bump, $M_\infty = 0.68$, 1095 Elements

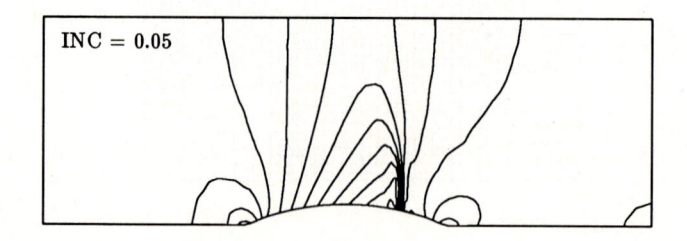

Figure 6.33: Pressure, Distorted Grid Solution for 10% Bump, $M_\infty = 0.68$

Chapter 7
Dispersion Phenomena and the Euler Equations

7.1 Introduction

In many solutions of the Euler equations, low wave number oscillations have been observed in the vicinity of rapid flow variations.[1] These oscillations cannot be explained by problems in artificial viscosity formulation, as their frequency is very low, and the amplitude is relatively insensitive to the amount of artificial dissipation used. Test cases computed for the author by his colleagues in the MIT Computational Fluid Dynamics Laboratory using methods ranging from cell-based finite volume schemes to flux vector split schemes all suffer from this oscillatory behavior. The cause of these oscillations is dispersion. This chapter analyzes the dispersive properties of the schemes for the solution of the Euler equations as presented in this thesis.

The approach taken here is to analyze the dispersive properties of the linearized, steady Euler equations on a regular mesh, using the spatial derivative operator for each of three methods discussed previously (Galerkin, cell-vertex, central difference). This analysis is applied to a model problem, and the prediction of the frequency and location of the dispersive oscillations is demonstrated. Finally, the theory is validated by comparison with numerical experiments.

7.2 Difference Stencils

Figure 7.1 shows the x difference stencils on a uniform mesh for the three methods. Some properties of the Galerkin and cell-vertex stencils are discussed further in the following sections.

---
[1] Kenneth Powell, Ken Morgan, Michael Giles, Robert Haimes, and others have observed these oscillations in various calculations.

```
   -1        1      -1       1       0        0
     ┌──┬──┐          ┌──┬──┐         ┌──┬──┐
   -4│  │  │4       -2│  │  │2      -1│  │  │1
     ├──┼──┤          ├──┼──┤         ├──┼──┤
     │  │  │          │  │  │         │  │  │
     └──┴──┘          └──┴──┘         └──┴──┘
   -1        1      -1       1       0        0
       Galerkin       Cell-Vertex     Central Difference
```

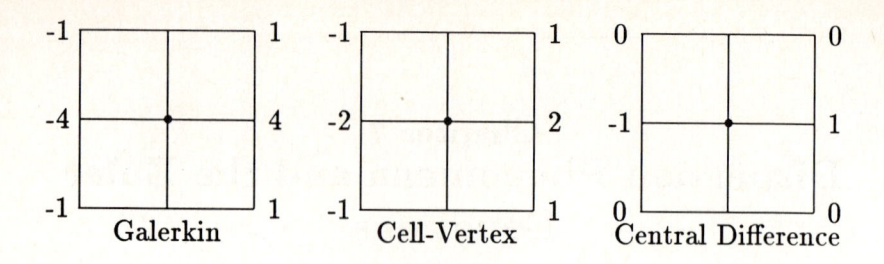

Figure 7.1: Difference Stencils for x Derivative, Three Methods

7.2.1 Some Properties of the Galerkin Stencil

This approximation has several interesting features. First, it gives the minimum steady-state error in the energy norm of the residuals, since the Galerkin method minimizes

$$\langle R, R \rangle \equiv \int R^2 dV, \tag{7.1}$$

where

$$R = \frac{\partial F}{\partial x} + \frac{\partial G}{\partial y}, \tag{7.2}$$

subject to the restriction that F and G are in the space spanned by the interpolation functions. Vichnevetsky and Bowles [74], and Strang and Fix [69] discuss this in more detail. Second, for the steady-state Euler equations on a uniform, parallelogram mesh, it is a fourth-order accurate approximation. This interesting feature was pointed out by Abarbanel [1] and can be seen by examining the truncation error for the x and y derivatives. The truncation errors for the derivatives of F and G with respect to x and y can be written (with $\mathcal{D}$ representing the derivative operator) as

$$\mathcal{D}_x F = F_x + F_{xxx}\frac{\Delta x^2}{6} + F_{xyy}\frac{\Delta y^2}{6} + \text{H.O.T.}, \tag{7.3}$$

$$\mathcal{D}_y G = G_y + G_{yyy}\frac{\Delta y^2}{6} + G_{xxy}\frac{\Delta x^2}{6} + \text{H.O.T.}, \tag{7.4}$$

so the discrete, steady residual $\mathcal{D}_x F + \mathcal{D}_y G$ can be written

$$\mathcal{D}_x F + \mathcal{D}_y G = F_x + G_y + \frac{\Delta x^2}{6}(F_{xxx} + G_{xxy}) + \frac{\Delta y^2}{6}(F_{xyy} + G_{yyy}) + \text{H.O.T.} \tag{7.5}$$

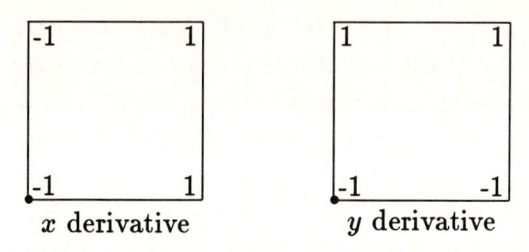

Figure 7.2: Difference Stencils For a Single Cell-Vertex Element

Now, note that in Eq. (7.5), the coefficients of the Δx^2 and Δy^2 terms are the derivatives of the quantity $F_x + G_y$ twice with respect to x and y. For the steady Euler equations, the quantity $F_x + G_y = 0$, so these terms will be higher-order. The transformation between the Cartesian (x, y) space and some other space in which the elements are parallelograms is a linear transformation, so all the results above also hold in the transformed space. Thus, on a regular mesh of parallelograms, the Galerkin method is fourth-order accurate. Note that this cancellation of truncation error fails if the equation is inhomogeneous (such as the Navier-Stokes or Conical Euler equations), or if the mesh is not composed of parallelograms (as in most practical problems).

7.2.2 Some Properties of the Cell-Vertex Stencil

The most interesting property of the cell-vertex scheme is that it is second-order accurate on any mesh of parallelograms, independent of mesh stretching. To see this, consider the contribution of a single element to the residual at a node. Figure 7.2 shows the difference stencils for the x and y derivatives at the node denoted by the dot.

Following the approach of the previous section, the truncation errors for the derivatives of F and G with respect to x and y can be written

$$\mathcal{D}_x F = F_x + F_{xx}\frac{\Delta x}{2} + F_{xy}\frac{\Delta y}{2} + \text{H.O.T.}, \tag{7.6}$$

$$\mathcal{D}_y G = G_y + G_{yy}\frac{\Delta y}{2} + G_{xy}\frac{\Delta x}{2} + \text{H.O.T.}, \tag{7.7}$$

so the discrete, steady residual $\mathcal{D}_x F + \mathcal{D}_y G$ can be written

$$\mathcal{D}_x F + \mathcal{D}_y G = F_x + G_y + \frac{\Delta x}{2}(F_{xx} + G_{xy}) + \frac{\Delta y}{2}(F_{xy} + G_{yy}) + \text{H.O.T.} \tag{7.8}$$

Again note that in Eq. (7.8), the coefficients of the Δx and Δy terms are the derivatives of the quantity $F_x + G_y$ with respect to x and y. For the steady Euler equations, the quantity $F_x + G_y = 0$, so these terms will be higher-order. Thus, on any mesh of parallelograms, the cell-vertex method should be second-order accurate. Note that this cancellation of truncation error again fails if the equation is inhomogeneous. This seems to indicate that the cell-vertex method may be superior on highly stretched grids, or on grids with embedded regions. For most practical problems, however, little difference is observed between the cell-vertex and Galerkin methods.

7.3 Linearization of the Equations

This section describes the linearizations of the Euler equations, using a method suggested to the author by Giles. The Euler equations (Eq. 2.1) can be rewritten in two dimensions as

$$\frac{\partial U}{\partial t} + A\frac{\partial U}{\partial x} + B\frac{\partial U}{\partial y} = 0, \qquad (7.9)$$

where

$$U = \begin{bmatrix} \rho \\ u \\ v \\ p \end{bmatrix}, \quad A = \begin{bmatrix} u & \rho & 0 & 0 \\ 0 & u & 0 & \frac{1}{\rho} \\ 0 & 0 & u & 0 \\ 0 & \gamma p & 0 & u \end{bmatrix}, \quad B = \begin{bmatrix} v & 0 & \rho & 0 \\ 0 & v & 0 & 0 \\ 0 & 0 & v & \frac{1}{\rho} \\ 0 & 0 & \gamma p & v \end{bmatrix}. \qquad (7.10)$$

The equations are linearized by "freezing" the A and B matrices. In the steady state, the time derivative vanishes, so one can write the linearized Euler equations in operator form as

$$(As_x + Bs_y)U = 0, \qquad (7.11)$$

where s_x and s_y are the x and y derivative operators. If non-trivial solutions to this equation are desired, the operator matrix $(As_x + Bs_y)$ must have zero determinant. This is the statement that

$$\begin{vmatrix} us_x + vs_y & \rho s_x & \rho s_y & 0 \\ 0 & us_x + vs_y & 0 & \frac{s_x}{\rho} \\ 0 & 0 & us_x + vs_y & \frac{s_y}{\rho} \\ 0 & \gamma p s_x & \gamma p s_y & us_x + vs_y \end{vmatrix} = 0. \qquad (7.12)$$

Define
$$r = \frac{s_x}{\sqrt{s_x^2 + s_y^2}}, \qquad (7.13)$$
$$s = \frac{s_y}{\sqrt{s_x^2 + s_y^2}}, \qquad (7.14)$$

and Eq. (7.12) can be expanded to

$$(ru + sv)^2 \left[a^2(r^2 + s^2) - (ru + sv)^2\right] = 0, \qquad (7.15)$$

where a is the speed of sound. This has solutions

$$ru + sv = \begin{cases} 0, \\ \pm a. \end{cases} \qquad (7.16)$$

Now, let

$$\vec{s} = \begin{bmatrix} r \\ s \end{bmatrix}, \qquad \vec{u} = \begin{bmatrix} u \\ v \end{bmatrix}, \qquad (7.17)$$

so that Equation (7.16) becomes

$$\vec{s} \cdot \vec{u} = \begin{cases} 0, \\ \pm a. \end{cases} \qquad (7.18)$$

Since $\vec{s}$ has unit norm, the non-zero solution will exist only if the flow is supersonic. So far, no restrictions have been placed on the derivative operators s_x and s_y. The analysis above applies to the exact derivative operators as well as any of the discrete operators. The next section introduces the discrete equations and their solution.

7.4 Fourier Analysis of the Linearized Equations

This section introduces the spatial discretizations of the equations into the linear model, and discusses the consequences of the truncation error in the approximations. Many of the ideas used here can be found in [74], but those analyses were performed for a scalar problem involving only one spatial direction and time.

For purposes of analysis, assume that the equations are discretized on a Cartesian $N_x \times N_y$ mesh with grid spacings in the x and y directions

Table 7.1: Spatial Derivative Operators for Various Methods

Method	s_x/i	$\mathcal{R}s_y/i$
Exact Derivative	ϕ	θ
Galerkin	$\frac{1}{3}\sin\phi(2+\cos\theta)$	$\frac{1}{3}\sin\theta(2+\cos\phi)$
Cell-Vertex	$\frac{1}{2}\sin\phi(1+\cos\theta)$	$\frac{1}{2}\sin\theta(1+\cos\phi)$
Central Difference	$\sin\phi$	$\sin\theta$

of Δx and Δy. Let $x = j\Delta x$ and $y = k\Delta y$, then assume the state vector is of the form

$$U(j\Delta x, k\Delta y) = \sum_{m=0}^{N_x-1}\sum_{n=0}^{N_y-1} \exp i(j\phi_m + k\theta_n)U'_{mn}, \qquad (7.19)$$

where ϕ_m and θ_n are spatial frequencies in the x and y directions and U'_{mn} is some eigenvector. The spatial frequencies are related to m and n by the relations

$$\phi_m = \frac{2\pi m}{N_x}, \qquad \theta_n = \frac{2\pi n}{N_y}. \qquad (7.20)$$

Next, consider a model problem in which $\Delta x = 1$, $\Delta y = \mathcal{R}$, $v \ll u$, and $u = Ma$. Equation (7.16) has the solution

$$\frac{s_x}{\sqrt{s_x^2 + s_y^2}} = \pm\frac{1}{M}. \qquad (7.21)$$

For a particular choice of spatial discretization, there is a particular dispersive character for a given Mach number M. Table 7.1 shows s_x and s_y for the Galerkin, cell-vertex and central difference methods, as well as the exact spatial derivative, assuming that ϕ and θ are continuous rather than discrete. Now introduce $s_1 = s_x$ and $s_2 = \mathcal{R}s_y$, square Eq. (7.21), and solve for s_1/s_2 to obtain

$$\frac{s_1}{s_2} = \mathcal{R}\sqrt{M^2 - 1}. \qquad (7.22)$$

This representation of the dispersion relation has the properties that s_1 and s_2 are functions only of the non-dimensional spatial frequencies ϕ and θ, and that all the problem- and grid-dependent terms are contained in the quantity $\mathcal{R}\sqrt{M^2 - 1}$, which will be called κ. Problems with similar values of κ should have similar dispersive behavior.

Dispersion brings in the concept of *group velocity*. A brief introduction to the concept of group velocity will be presented here. For a more detailed discussion, any wave mechanics text (such as [14]) should be sufficient. Consider a one dimensional wavelike disturbance,

$$q(x,t) = e^{i(kx-\omega t)} \;, \qquad (7.23)$$

where k is the wave number and ω is the temporal frequency of the wave. The quantity ω/k represents a velocity at which wave "crests" travel, and is called the *phase velocity*. Without going into the mathematical details, the quantity $d\omega/dk$ is another velocity, the *group velocity*, and represents the velocity at which a wave packet travels. For example, deep water waves have a phase velocity which is twice the group velocity. This means that if one observes a group of deep water waves, the crests will appear to move faster than the group of waves. In this analysis, identify y with t so that y is treated like "time," then ϕ is like k and θ is like ω. Now one can interpret plots of θ vs. ϕ. The slope of a curve on which κ is constant is $d\theta/d\phi$, which is the "spatial group velocity", or the *angle* at which wave packets propagate. Waves with large spatial group velocity (the angle on the θ/ϕ plot is close to vertical) will travel at a shallow spatial angle (the wave will move a long way in x for a little change in y). This allows one to predict where the dispersed waves will appear. Figure 7.3 shows the contours of constant κ for the exact spatial derivative operator. Note that the lines of constant κ are straight, indicating that all frequencies travel at the same angle. Moreover, for $\mathcal{R} = 1$, the waves have angle $\tan^{-1}(1/\sqrt{M^2-1}) = \sin^{-1}(1/M)$, which is just the Mach angle. This is expected since the linearized Euler equations were considered.

Figure 7.4 shows the contours for the Galerkin method. Note that the curves are multiple-valued. For a given κ and ϕ there may be more than one value of θ, or no values of θ. The second branch is at a high spatial frequency, which will be attenuated by the artificial viscosity in the scheme, so this branch does not affect the solutions. The other noteworthy thing about Fig. 7.4 is that the curves depart from the exact Euler curves much later than all the other methods. This is due to the fact that on a uniform, parallelogram mesh, the Galerkin method is fourth-order accurate for the linearized Euler equations. In practice, one never sees this fourth-order accuracy, for three reasons. First, the artificial viscosity introduces some error into the solution scheme. Lindquist has shown [38] that artificial viscosity can have a dominant effect on the solution error. Second, the grids used to solve problems are very seldom uniform parallelograms. On a mesh composed of anything other than congruent parallelograms, fourth-order accuracy is no longer obtained. Third, the flux calculations for the spatial discretization introduce some error. Since the fluxes are non-linear functions of the state variables, the assumption that both the state vectors and flux vectors vary linearly (or quadratically) within the element introduces an error. Roe [61] has shown that these errors are usually second order. These effects combine to make the Galerkin scheme second-order accurate for practical problems.

Figure 7.5 shows the dispersion plot for the central difference method. Note that the character of the diagram is similar to the Galerkin plot. Also note that the curves depart from the exact curves at a lower frequency than the Galerkin curves in Fig. 7.4. One would expect the dispersive behavior to be similar to the Galerkin dispersive behavior, and to some extent, this is the case.

Figure 7.6 shows the dispersion curves for the cell-vertex scheme. Note that the curves are single-valued. Also note that the curvature is opposite the curvature for the Galerkin and central difference methods. For a particular choice of κ, the dispersion curve for the cell-vertex method will lie on the opposite side of the exact dispersion line than the curves for the Galerkin and central difference methods. This implies that the oscillations due to dispersion at a feature (a shock, for example) should appear on the opposite side (ahead or behind) of the feature compared to the Galerkin and central difference oscillations.

An important application of these curves is the prediction of oscillations due to discontinuities such as shocks. In some problems, oscillations before or after a shock can cause the solution algorithm to diverge. For example, in a strong expansion, a post-expansion oscillation may drive the pressure negative, while a pre-expansion oscillation may not be harmful. The dispersion curves allow one to predict the location of these oscillations and choose a solution algorithm which will put them in a safe place. The location of oscillations may be predicted by the following rule: If the θ vs. ϕ curve is concave up, the oscillations will be behind the feature (they travel faster than the exact solution), and if the curve is concave down, the oscillations will be ahead of the feature. For the cell-vertex method, this means that one will see pre-feature oscillations for $\kappa > 1$ and post-feature oscillations for $\kappa < 1$. For the Galerkin and central difference methods, this is reversed: $\kappa < 1$ implies pre-feature oscillations, and $\kappa > 1$ implies post-feature oscillations.

7.5 Numerical Verification

To verify that the theory above is the correct explanation of the oscillations observed in many problems, several numerical experiments were made. All test problems were for flow over a wedge in a channel with various wedge angles, inflow Mach numbers and mesh aspect ratios. Figure 7.7 shows the geometry and flow topology for the 10° wedge, $M_\infty = 2$ case discussed below. All the calculations were performed on 50x20 grids, and result in similar flow topologies.

The first set of experiments is for a 1/2° wedge angle, with $\kappa = 1.732$. Figure 7.8 shows the curves for all three numerical methods and the exact spatial derivatives on a single plot for this value of κ. Here it is apparent that the Galerkin curve stays much closer to the exact curve. Three numerical test cases were run: Mach 2 flow with $\mathcal{R} = 1$; Mach 1.323 flow with $\mathcal{R} = 2$; and Mach 3.606 flow with $\mathcal{R} = 1/2$. A quick examination of the flow geometry gives the physical significance of κ as the ratio of the number of x grid lines crossed by the feature per y grid line crossed. In the Mach 3.606 flow, the shock lies at a much shallower angle, so that for a smaller Δy the same crossing ratio is obtained. A similar argument applies to the Mach 1.323 flow.

Figure 7.9 shows the Mach number at mid-channel for the central difference method, scaled by the free stream Mach number for the different Mach numbers above. The central difference method is used here because it exhibits the most oscillation with the greatest amplitude. The exact Mach number ratios (M/M_∞) for these shocks are 0.991 for $M_\infty = 2$ and $M_\infty = 3.606$ and 0.986 for $M_\infty = 1.323$. These compare well with the actual data, and explain why two of the curves lie on top of each other. Note that the frequencies of the oscillation are nearly identical. Also note that the frequency changes slightly as one moves further downstream of the shock. This is as predicted by the dispersion curve. As one moves downstream, the spatial group velocity increases, meaning ϕ increases slightly. The wavelength predicted by the dispersion relation at $(x,y) = (1.5, 0.5)$ should be about 10.5 points, and the measured wavelength (crest-to-crest) is either 10 or 11 points, depending on where one defines the crest.

The next set of data shows the location of the oscillations for the Mach number 2 case with the three methods. In all the figures shown, the plot is of Mach number at mid-channel. Figure 7.10 shows the plot for the Galerkin method, Figure 7.11 for the central difference method and Figure 7.12 for the cell-vertex method. Note that both the Galerkin and central difference methods exhibit post-shock oscillation, while the cell-vertex exhibits pre-shock oscillation. Also note that the frequency of the Galerkin oscillations is much higher, and has a lower amplitude than the oscillations from central difference approximation. This is expected since the Galerkin method group velocity errors occur at higher spatial frequencies (Fig. 7.8). As an interesting aside, note that in Fig. 7.12 the pre-shock oscillations from the reflected shock are visible at the right side of the plot. These figures verify the use of the dispersion curves to predict the location of dispersive phenomena.

The final test cases show the Mach 2 flow over a 10° wedge, which generates a shock wave with a normal Mach number of 1.27 and a density ratio of 1.46. This case was chosen because the problem starts to become significantly nonlinear. Figure 7.13 shows Mach number on a slice at a 5° angle to the x axis for the Galerkin method and Fig. 7.14 shows the same data for the cell-vertex method. In these cases, κ is about 1.7 ahead of the first shock, and κ is about 0.6 behind the reflected shock (due to the lower Mach number of 1.28 and the changing aspect ratio of the cells). In between the shocks, κ is about 1.04. Note that the oscillation positions in Fig. 7.14 are correctly predicted to be before the first shock, where $\kappa > 1$, and after the second (reflected) shock, where the $\kappa < 1$. Note also that the frequency of the oscillations has increased. This may be due to the tendency of nonlinear shocks to self-steepen. This self-steepening behavior has almost completely eliminated the dispersive error in the Galerkin case, but in Fig. 7.13 small low-frequency oscillations are still visible after the first shock and before the second shock. In many practical applications, the analyst is concerned with the location (pre- or post-shock) more than the frequency, and the linearized analysis is still useful for those applications.

7.6 Conclusions

The primary conclusion of this study is that the low frequency oscillations sometimes seen near shocks and other discontinuities are due to dispersion in the numerical scheme. The linearized analysis presented gives one a method for predicting the location and frequency of these oscillations. The linear analysis is also effective in predicting the location of the oscillations for problems with significant nonlinearity. The central difference finite element method is demonstrated to be inferior to the Galerkin and cell-vertex methods due to its poor dispersive behavior. The Galerkin finite element is fourth-order accurate for uniform meshes and has the lowest dispersive error. The cell-vertex method is second-order accurate for any parallelogram mesh and has moderate dispersive error. Either of the Galerkin or cell-vertex methods provides acceptable dispersive performance.

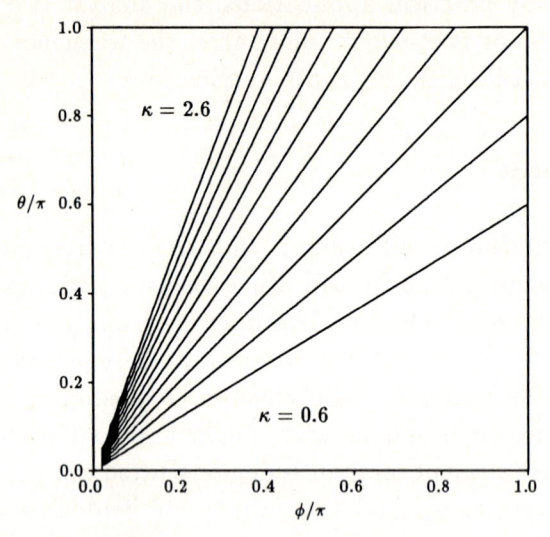

Figure 7.3: Lines of Constant κ for Exact Spatial Derivatives

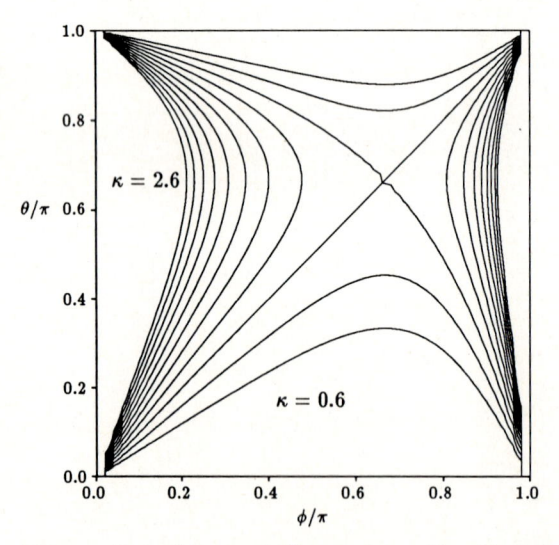

Figure 7.4: Lines of Constant κ for Galerkin Method

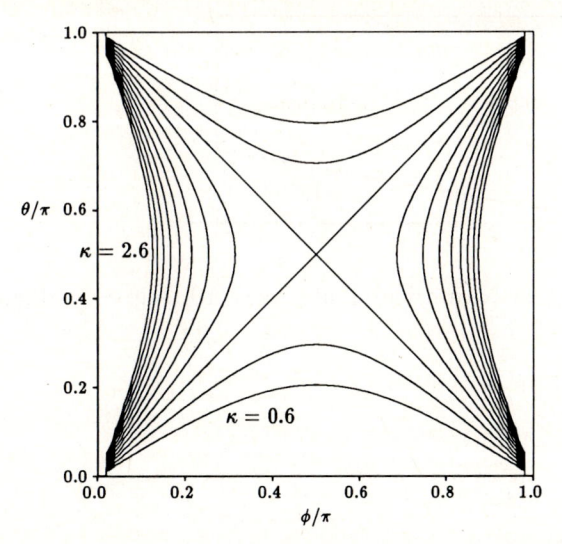

Figure 7.5: Lines of Constant κ for Central Difference Method

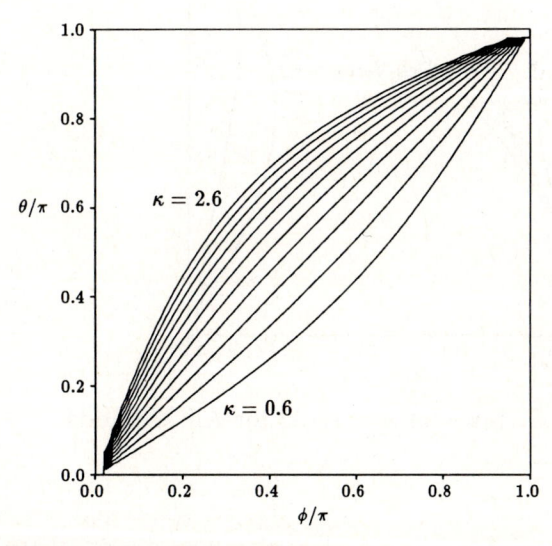

Figure 7.6: Lines of Constant κ for Cell-Vertex Method

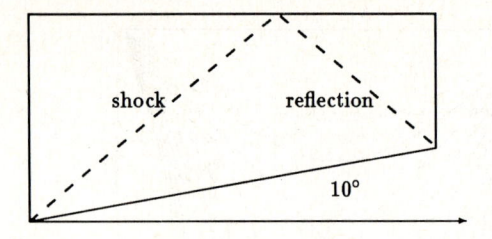

Figure 7.7: Geometry for Numerical Test Cases, 10 Degree wedge, $\mathcal{R} \approx 1$

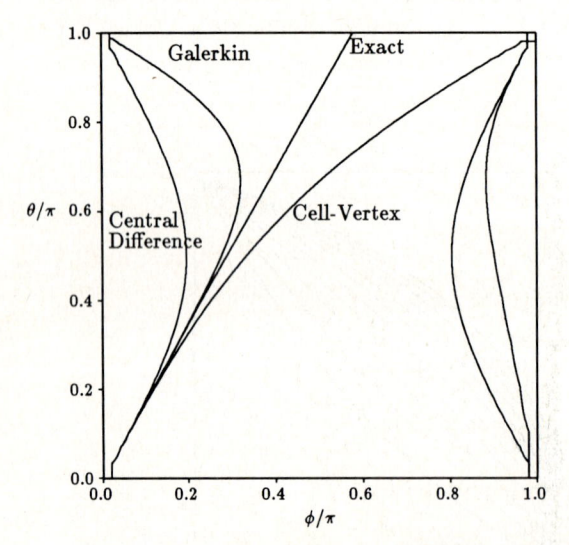

Figure 7.8: Lines of $\kappa = 1.732$ for All Methods

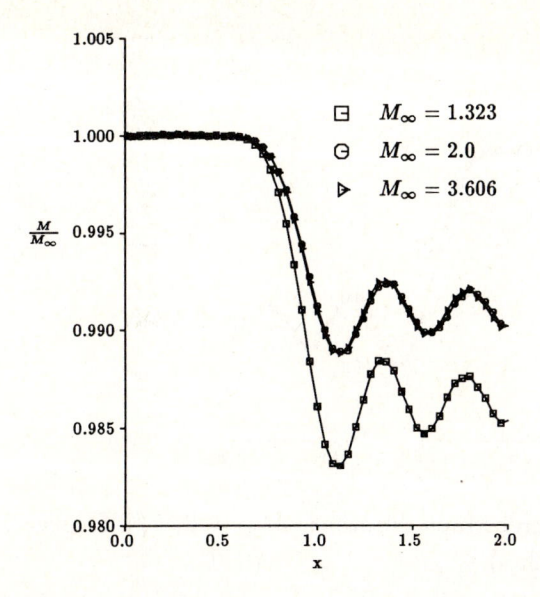

Figure 7.9: M/M_∞ for $\kappa = 1.732$, Central Difference Method

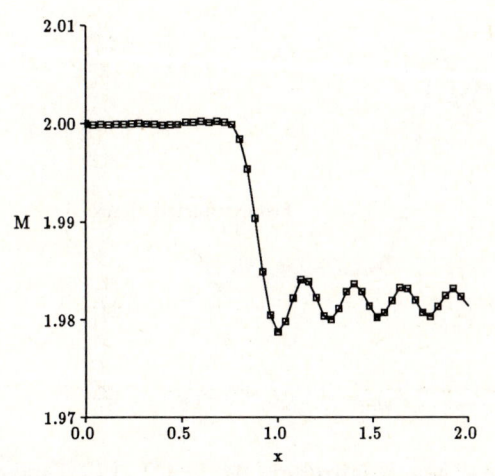

Figure 7.10: Mid-channel Mach Number, $M_\infty = 2$, 1/2 Degree Wedge, Galerkin Method

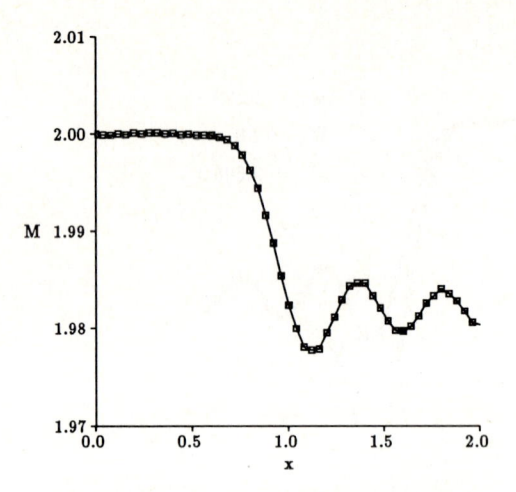

Figure 7.11: Mid-channel Mach Number, $M_\infty = 2$, 1/2 Degree Wedge, Central Difference Method

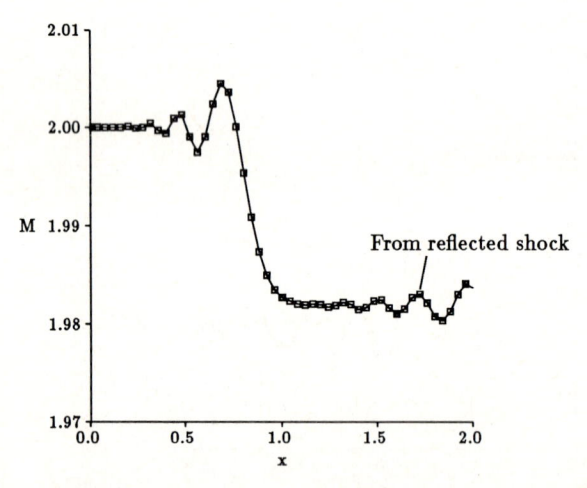

Figure 7.12: Mid-channel Mach Number, $M_\infty = 2$, 1/2 Degree Wedge, Cell-Vertex Method

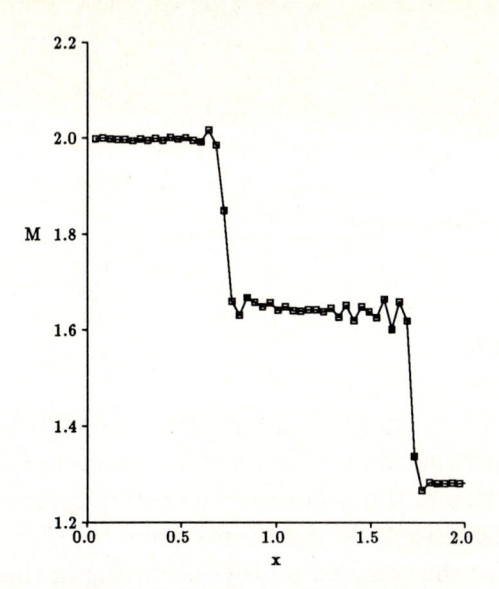

Figure 7.13: Mid-channel Mach Number, $M_\infty = 2$, 10 Degree Wedge, Galerkin Method

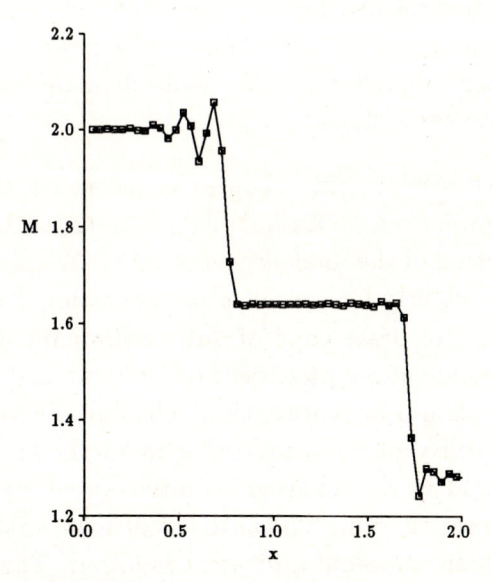

Figure 7.14: Mid-channel Mach Number, $M_\infty = 2$, 10 Degree Wedge, Cell-Vertex Method

Chapter 8
Scramjet Inlets

8.1 Introduction

The current National Aerospace Plane program (NASP) has sparked renewed interest in hypersonic flow. One particular aspect of this project attracting much attention is the proposed supersonic combustion ramjet or *scramjet* propulsion system. Scramjets pose an interesting (and difficult) problem due to the complex physics occurring in the engine. A survey article by White [75] discusses some of the issues involved in computing scramjet flows. This chapter focuses on the flows in the scramjet inlets. Although the physical model used here (the Euler equations) lacks many of the features needed for anything but preliminary design (such as viscosity, chemistry and heat transfer), the physics of flows in scramjets is complicated enough that the inviscid model can provide some useful insights into these flows.

The model geometry used in this chapter is based on a two-strut design published in a paper by A. Kumar [31]. The three dimensional design is a modified version of the models presented in [32] and [33]. The geometry was modified slightly for ease of grid generation. No attempt was made to produce an accurate copy of any preliminary design, but merely to demonstrate some of the physics of these flows, and the ability of the finite element algorithm to capture these physics. These examples also demonstrate the utility of unstructured grid methods. The two-dimensional scramjet grid in this chapter required about two hours of *real* time to generate, from the time this author sat down with Kumar's paper to the time the computational grid was generated. The extension of this grid to include the inflow region only took an additional half hour. The computer code used to generate these test results is identical to the code used in the previous chapters to generate the bump and wedge solutions. The three-dimensional grid required about two hours

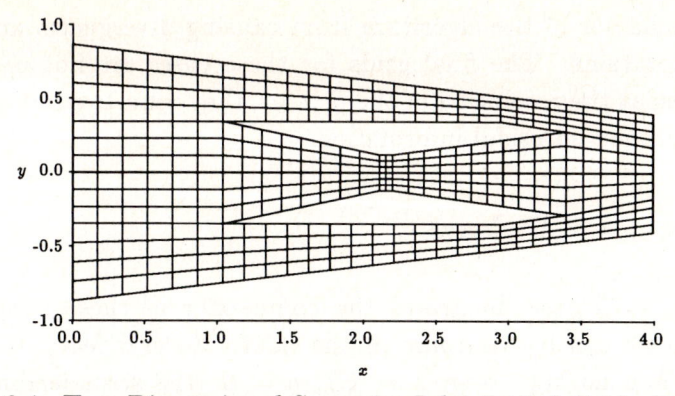

Figure 8.1: Two-Dimensional Scramjet Inlet Initial Grid, 416 Elements

to generate, not counting the time it took to debug the grid generator program. For reference, the complete geometry specification for the two and three dimensional models is presented in appendix B.

8.2 Two-Dimensional Test Cases

Several representative test cases were computed in two dimensions. The cases were chosen because they illustrate a wide variety of intersting physical features. Figure 8.1 shows the geometry and initial grid for all test cases. This geometry is a slice of the three-dimensional inlet taken roughly parallel to the wings, and the two struts in the middle are used to inject fuel. The initial compression surfaces slope at 6.7° and the strut compression surfaces slope at 11.9°.

The test conditions used were $M_\infty = 5$, $M_\infty = 3$, and $M_\infty = 2$ at 0° yaw, $M_\infty = 5$ at 5° yaw, and $M_\infty = 3$ at 7° yaw. Since the slice is taken parallel to the wings, the angle of yaw corresponds to the flow angle at the inlet. The $M_\infty = 2$ case was chosen to illustrate choked flow, and the $M_\infty = 3$ cases were chosen because they illustrate a situation in which yaw causes the flow to choke, unstarting the inlet. The $M_\infty = 5$ cases demonstrate that the yaw problem decreases at higher Mach numbers. All of these except the $M_\infty = 3$, 7° angle of attack case were computed using the cell-vertex method, with 4 levels of grid embedding and 5 adaptations using the second difference indicator. The $M_\infty = 3$, 7° angle of attack case required special treatment to prevent the pseudo-unsteady behavior of the algorithm from causing divergence, and used only 2 adaptations. The final grids for these cases are not presented here because at the resolution of the printer, they tend to print as black blobs and convey no useful information.

8.2.1 $M_\infty = 5$, 0° Yaw

The $M_\infty = 5$ case illustrates the complexity of these inlet flows. Fig. 8.3 shows density contours in the inlet. Several features of this flow are worth noting. Near $x = 1.7$, $y = 0$ the strut leading edge shocks intersect, and both bend toward the flow direction. Notice the two expansion fans from the flat throat area near $x = 2.1$, and observe the bending of the strut shocks toward the walls as they pass through the expansion. In particular, note that the reflection of the strut shock curves toward the centerline, and that the curvature changes as the downwind expansion fan interacts with the shock. Following along the shock, notice that at $x = 2.7$, $y = -0.1$, the expansion fans from the top wall begin to weaken this shock, so that near $x = 3$, the shocks have practically vanished. Shifting to the side passages, the leading edge shock reflects twice, strengthening (and steepening) each time, and then near $x = 3$ passes through an expansion fan and weakens. Finally, the shock strikes the trailing edge and reflects out of the domain, with the reflection of the expansion fan weakening the shock still further as it goes. A consequence of the differences in flow between the two passages is that a slip line and shock form off the strut trailing edge. Returning to the central passage, the very weak strut leading edge shock strikes the strut trailing edge shock, causing both to bend slightly and strengthening the very weak shock. As Fig. 8.4 shows, the pressure contours tell a similar tale, but note that the slip line off the trailing edge does not appear. This case has 12998 elements and required 37 minutes to compute on the Alliant.

These same features can also be seen in the unadapted biquadratic test case in Fig. 8.5. This case was computed using 416 biquadratic elements (the equivalent of the initial bilinear mesh). The resolution of shocks is poorer, but this is to be expected since the equivalent globally fine bilinear mesh for these cases would have 26624 elements. Also, this biquadratic test case only took 1.25 minutes to compute on the Alliant, for a savings of a factor of 30. To be fair to the bilinear elements, a 416 element biquadratic mesh gives approximately the same accuracy as a 6656 element bilinear mesh, which requires about 8 minutes to compute, reducing these savings to a factor of 6.

8.2.2 $M_\infty = 5$, 5° Yaw

The next case examined was the same inlet with $M_\infty = 5$ at 5° yaw. Contours of pressure are shown in Fig. 8.6. There are several differences between this flow and the previous flow. First, note the presence of two shocks on the lower strut and of a shock and an expansion on the upper strut. In particular, note the interaction between the strut expansion and the inlet leading edge shock near $x = 1.5$. Second, follow the inlet leading edge shock through this expansion, then through the reflection of the expansion, then through the expansion at $x = 2.8$, then finally through the reflection of this expansion and out the domain. The reflected strut leading edge expansion is visible (barely) as far downstream as $x = 3$ (Fig. 8.7). Third, note the differences in bending between the two strut shocks in the central passage, and note that the shock from the lower strut weakens and nearly vanishes by the time it reaches the exit, while the shock from the upper strut is strengthened by the passage through the trailing edge shock near $x = 3.6$. A close examination of this shock reveals a mysterious bending and strengthening of this shock near $x = 3.8$. This is not mysterious when one looks at the density contours in Fig. 8.7, which clearly show the interaction of the slip line and the shock. This close-up also shows a weak expansion fan coming off the trailing edge between the slip line and the third reflection of the inlet leading edge shock, due to the difference in angle between the slip surface and the upper surface of the strut. This expansion further weakens the reflected inlet leading edge shock.

8.2.3 $M_\infty = 2$, 0° Yaw

The $M_\infty = 2$ test case is interesting because the flow chokes in the inlet, resulting in a solution that depends on the exit pressure. This choking is highly undesirable, as the engine would "unstart." In this case, the exit pressure was arbitrarily set to 2.5, corresponding to 3.5 times free-stream pressure. In order to correctly compute the flow, the computational grid must be extended in front of the inlet to capture the bow shock. No adaptation was allowed at the bow shock, since the region of interest is inside the inlet. Figure 8.8 shows the density contours in the inlet. Note the presence of the strong normal shock ($M_n = 1.8$) in the center passage. The differing flow conditions in the passages produce a slip line, as seen in the figure. The pressure contours in Fig. 8.9 shows the same flow features, with the slip line clearly absent. Finally, Fig. 8.10 shows the Mach number in the middle of the channel. Note that the Mach number is 1 at the throat, as expected for choked flow. This case used 2568 elements and required 9.5 minutes of Alliant CPU time.

8.2.4 $M_\infty = 3$, 0° Yaw

The flow topology for the $M_\infty = 3$ case is shown in the density contours of Fig. 8.11. There are several interesting features to observe here. Note that the shock off the inlet entrance reflects six times, and in particular note that near $x = 3.7$, $y = 0.3$, the shock bends slightly toward the center line. Now look at the strut leading edge shocks, and note that the third reflection of the shocks (starting near $x = 2.8$, $y = \pm 0.3$, and barely visible as "dents" in the contour lines) passes through the sixth reflection of the entrance shock where the entrance shock experiences the bend. Another cause of the bending is the interaction with the slip line off the trailing edge. This slip line is not readily apparent in Fig. 8.11, but the plot of total pressure loss at the exit in Fig. 8.12 clearly indicates its presence. Note the jumps due to the reflected inlet entrance shock and the trailing edge shocks as well.

Another interesting portion of the flow, highlighted in the Mach number contours of Fig. 8.13, is the flow near the throat. In this region, the strut shocks interact with each other and with the expansion fans off the throat in a very small region. An interesting characteristic of this interaction is that, at the intersection of the two strut shocks, the flow reaches Mach 1. If the Mach number is reduced further, a small region of subsonic flow is produced behind these shocks, and at about Mach

2.85, the center passage chokes and the inlet unstarts. Figure 8.14 shows the Mach number in the center of the channel, verifying that the Mach number drops to 1 just before the throat. Also note that in this figure, near $x = 3.9$, two shock jumps are visible, corresponding to the trailing edge strut shocks and the inlet entrance shocks, respectively.

The biquadratic elements were also applied to this case, and density contours are shown in Fig. 8.15. The resolution is not as good as Fig. 8.11, but the bilinear case used 14648 elements and required 51 minutes to compute, while the biquadratic case used 416 elements and required 2 minutes to compute. This case further verifies the usefulness of the biquadratic elements.

8.2.5 $M_\infty = 3$, 7° Yaw

The final two-dimensional problem examined is the inlet with $M_\infty = 3$ at 7° yaw. Under these conditions, the inlet actually unstarts. Figure 8.16 shows the Mach number in the unstarted inlet. As in the $M_\infty = 2$ case, the grid must be extended in order to capture the bow shock. In this case the exit pressure was set to 5.75, corresponding to 8 times the free stream pressure. The flow topology is very similar to the flow topology in the $M_\infty = 2$ case above, but there is asymmetry due to the angle of attack. As Fig. 8.16 shows, the bow shock has a large amount of asymmetry, while the flow in the inlet is only slightly asymmetrical. Figure 8.17 shows the Mach numbers in the upper, middle and lower channels. The asymmetry is so small in the inlet because most of the asymmetrical effects are buffered out in the bow shock. Coincidentally, the mid-channel shock has $M_n = 1.8$ in this case as well.

8.2.6 Inlet Performance and Total Pressure Loss

An important measure of inlet performance is the total pressure loss in the inlet. The scramjet inlets calculated above all have significant losses due to the compressions through the multiple shock systems. It is desirable to design the inlet so that these shock losses are minimized at the cruising Mach number, while still maintaining adequate loss at off-design points. Figure 8.18 shows the mid-channel total pressure loss

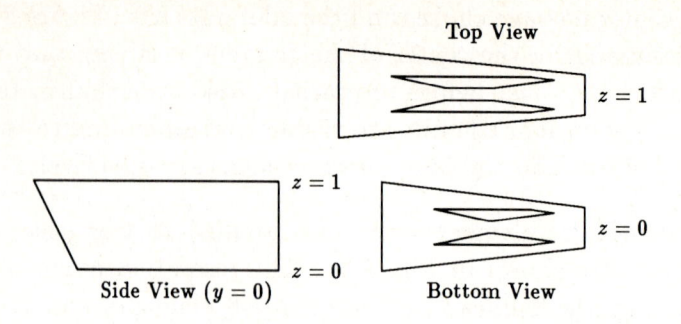

Figure 8.2: Geometry of Three-Dimensional Inlet

in the inlet for the five cases discussed above. This figure indicates the very large losses which occur when the inlet unstarts. For example, in the $M_\infty = 3$, $\alpha = 0°$ case, the maximum total pressure loss in the mid-channel is only 26%, while in the $M_\infty = 3$, $\alpha = 7°$ case, the mid-channel total pressure is 75%. The negative undershoots occuring at the shocks are due to numerical error, and have no physical significance. Viscosity and chemistry also play major roles in the engine, limiting the applicability of the inviscid simulation. However, the inviscid simulations presented here allow the designer to get a rough idea of the flow physics, and provide insight into the potential problems associated with these engines.

8.3 Three-Dimensional Results

A three-dimensional test case with $M_\infty = 5$ was computed to examine the effects of the third dimension on these flows. The geometry of the three-dimensional inlet is shown in Fig. 8.2, and is based on the two-dimensional geometry above. The leading edge sweep angle is 30°. A slice at $z = 0$ is identical to the two-dimensional geometry, and the slice at $z = 1$ has the portion of the struts and compression surfaces forward of the throat extended to give a 30° leading edge sweep. This weakens the compressions near $z = 1$, and forces the flow to turn down. Although the actual inlet proposed for the NASP project has a cowl plate which extends back from the throat, for the purposes of this study the cowl plate will be assumed to extend to the inlet mouth. For this case, contours of density at three z locations are shown in Fig. 8.19. Compare the 3-D contours with the 2-D, $M_\infty = 5$ density contours shown in Fig. 8.3, and note the differences introduced by the third dimension. Unfortunately, due to problems with the adaptation indicator, the side passages did not receive enough adaptation in the 3-D case, but the central passage indicates many of the important differences. At $z = 0.5$ and $z = 0.87$, note that that the strut leading edge shocks reflect multiple times in the throat area, resulting in a significantly different flow in the expansion region. Also note the differences in the slip line positions for the 3 z slices. in the $z = 0.13$ slice, the slip line and the trailing edge shock are nearly on top of one another, while at $z = 0.87$ they are quite distinct. Figure 8.20 shows a density slice at $y = 0$. Features to note here are the coalescences of the multiple reflections in the throat area near $x = 1.2, z = 0.2$, and also a very faint, 3-D reflection extending from about $x = 1, z = 0$ to about $x = 2, z = 1$, due to the turning produced by the 30° strut sweep. This case was computed adaptively on the IBM 3090VF at Cornell University, had 188692 elements on the final grid and took over 12 hours of CPU time. Unfortunately, the mesh is not fine enough to capture many of the interesting interactions. In order to obtail the resolution equivalent to the resolution in the two-dimensional cases above, over 1.3 million elements would be needed, requiring about 480 megabytes of memory and about 1 week of single-processor IBM 3090 CPU time.

A test case with $M_\infty = 4$ was also computed using the MIT Cray II. This case used 269826 nodes and 218554 elements, and required about 4 hours of CPU time. Contours of pressure at $z = 0.5$ are shown in Fig. 8.21, and contours of density at the same location are presented in Fig. 8.22. The basic features of the flow for $M_\infty = 4$ are similar to the features for $M_\infty = 5$, but note that in the $M_\infty = 4$ case, the shocks are a little steeper. This case also illustrates the need for large numbers of mesh points (even with adaptation) for problems involving complex, 3-D shock interactions.

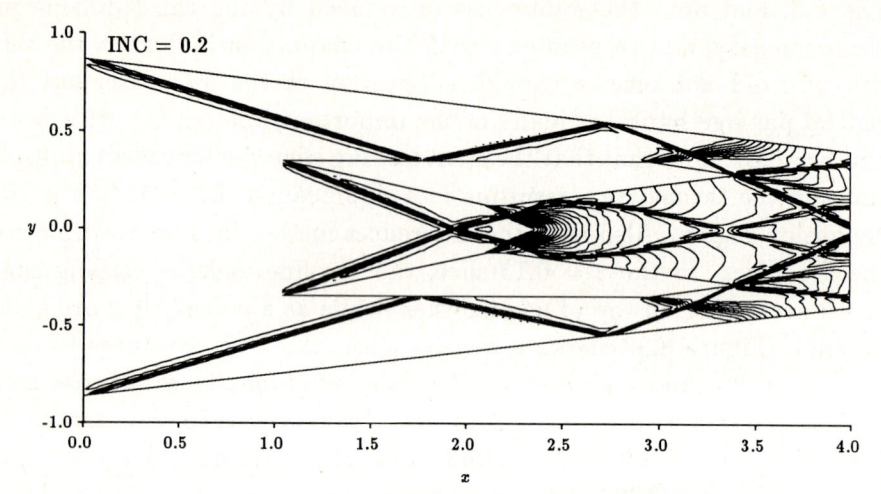

Figure 8.3: Density, $M_\infty = 5$, $\alpha = 0°$

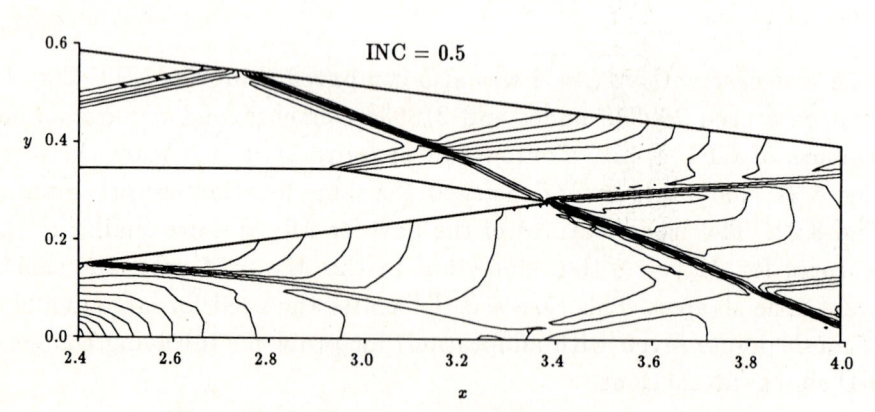

Figure 8.4: Pressure close-up, $M_\infty = 5$, $\alpha = 0°$

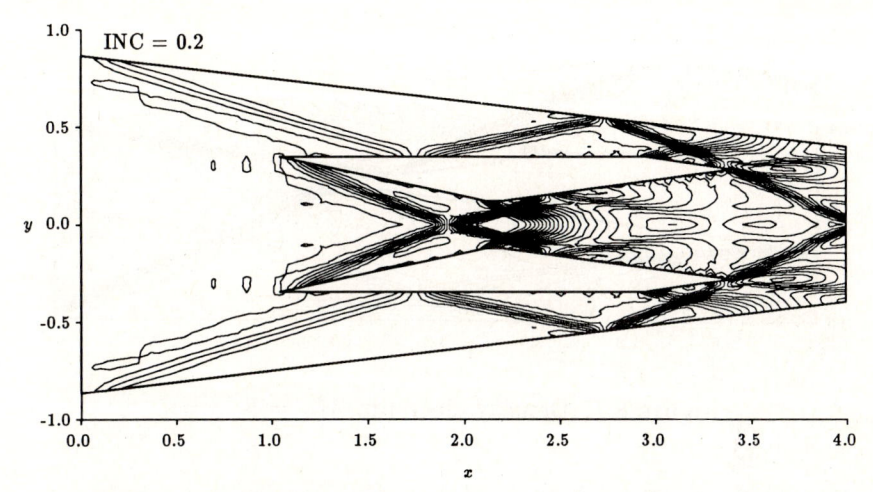

Figure 8.5: Density, $M_\infty = 5$, $\alpha = 0°$, biquadratic elements

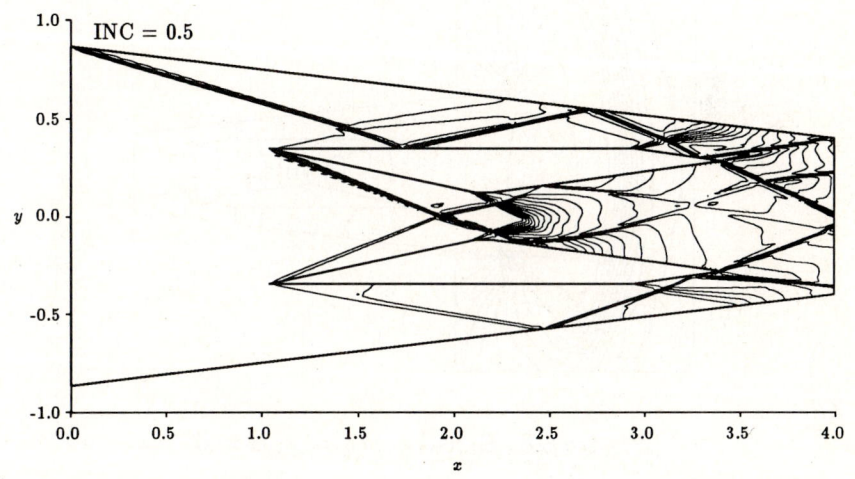

Figure 8.6: Pressure, $M_\infty = 5$, $\alpha = 5°$

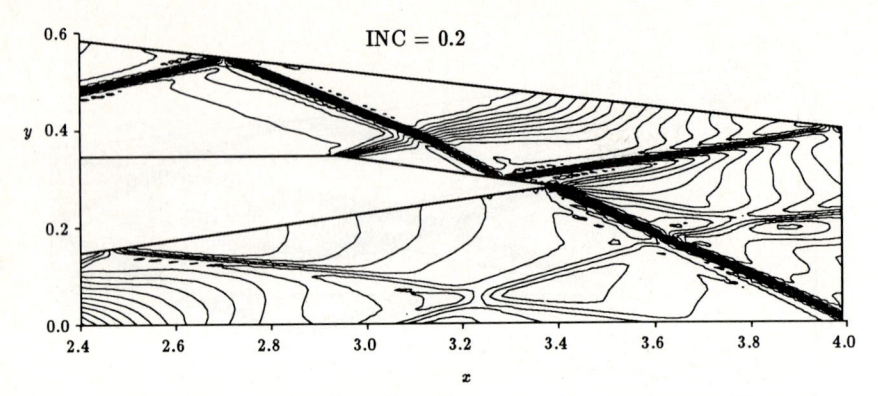

Figure 8.7: Density close-up, $M_\infty = 5$, $\alpha = 5°$

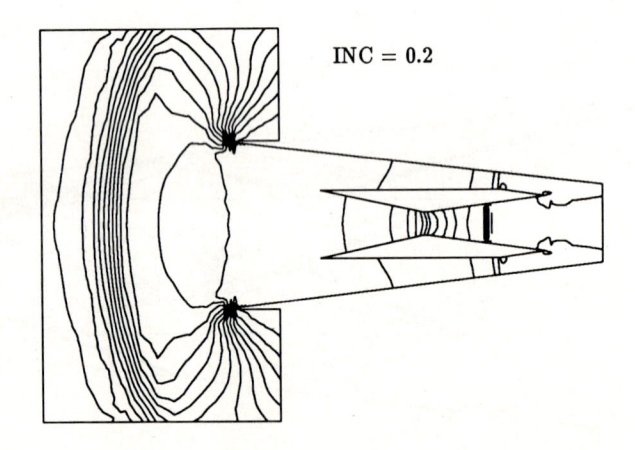

Figure 8.8: Density, $M_\infty = 2$, $\alpha = 0°$

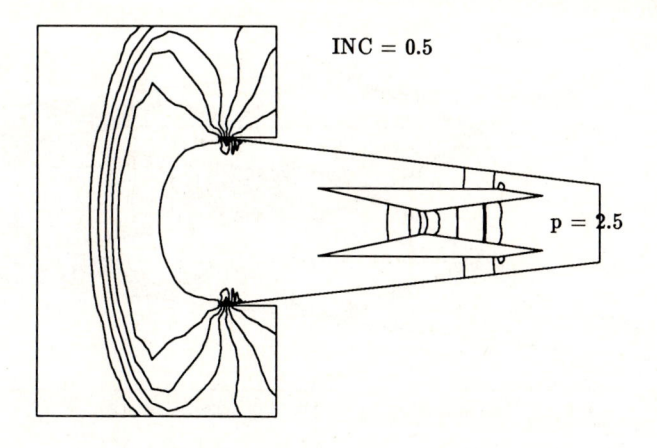

Figure 8.9: Pressure, $M_\infty = 2$, $\alpha = 0°$

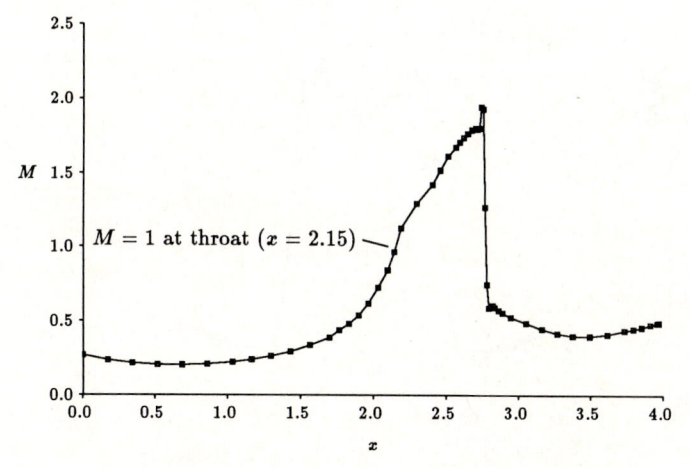

Figure 8.10: Mid-channel Mach number, $M_\infty = 2$, $\alpha = 0°$

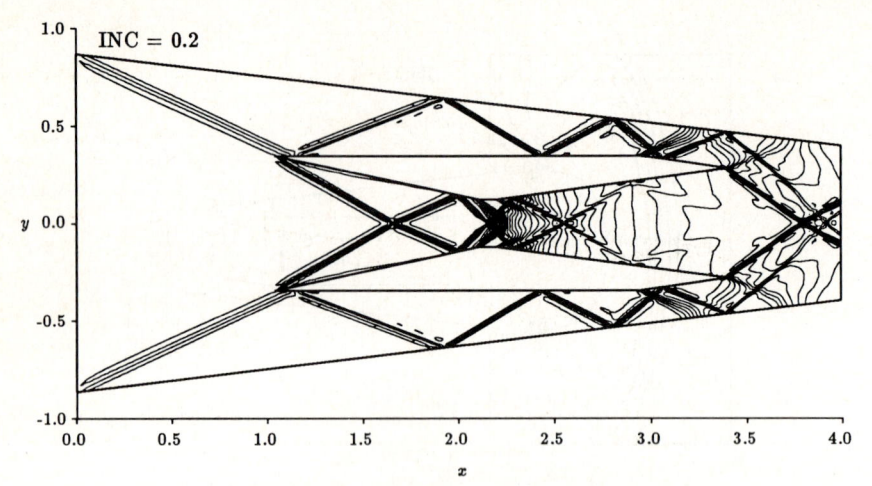

Figure 8.11: Density, $M_\infty = 3$, $\alpha = 0°$

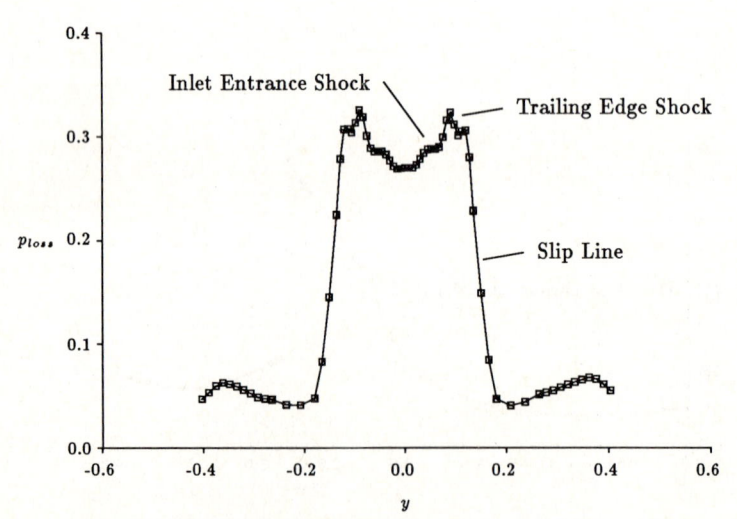

Figure 8.12: Total pressure loss at exit, $M_\infty = 3$, $\alpha = 0°$

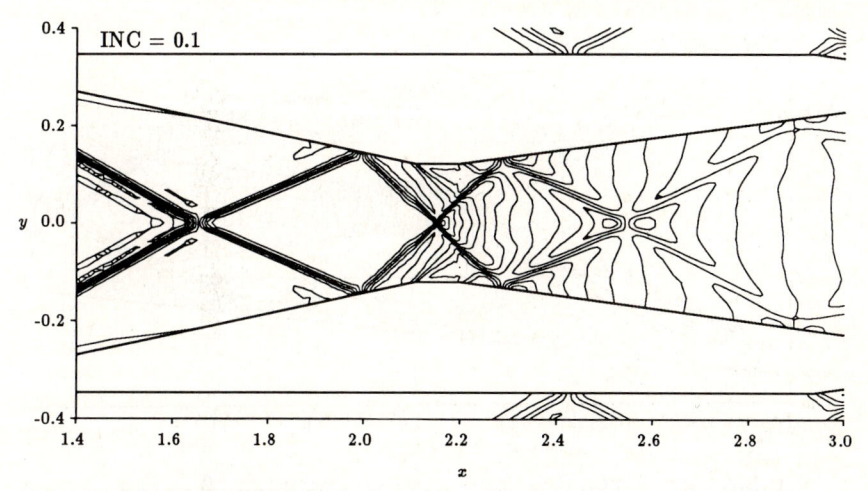

Figure 8.13: Mach number close-up, $M_\infty = 3$, $\alpha = 0°$

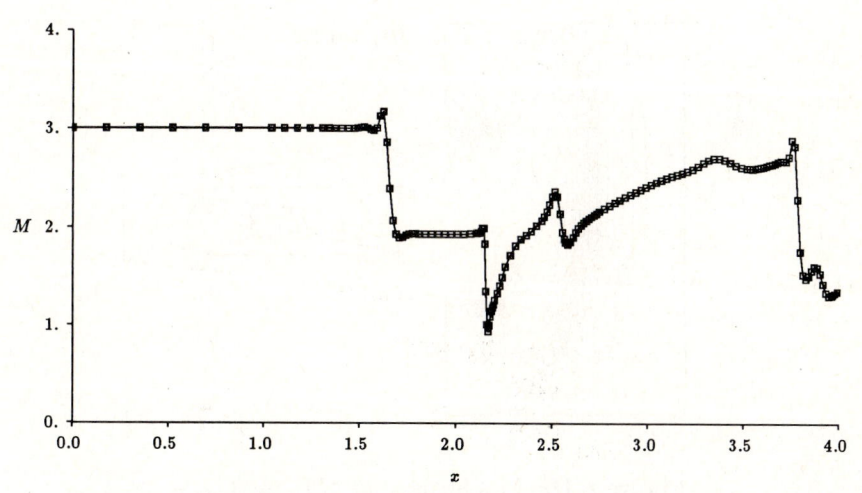

Figure 8.14: Mid-channel Mach number, $M_\infty = 3$, $\alpha = 0°$

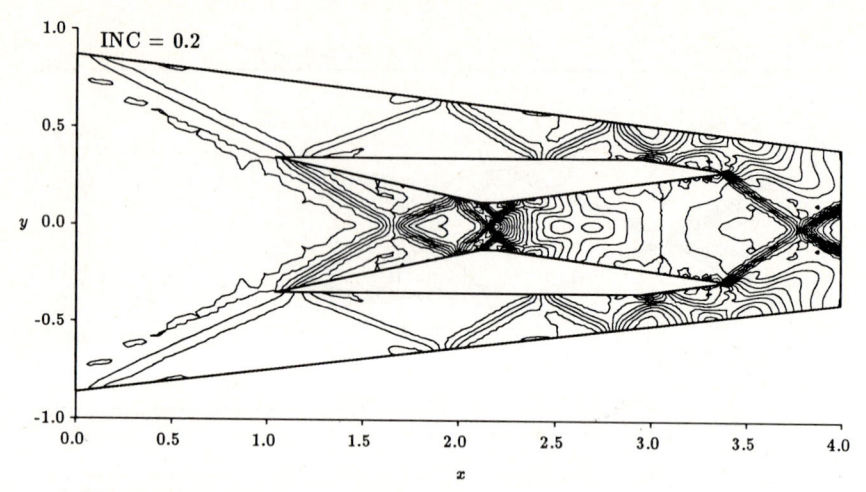

Figure 8.15: Density, $M_\infty = 3$, $\alpha = 0°$, biquadratic elements

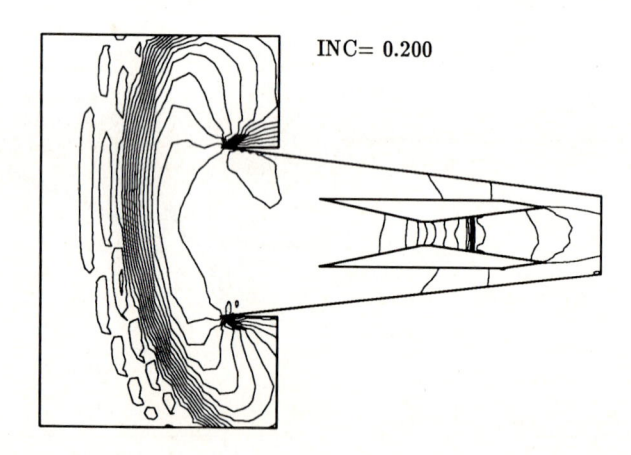

Figure 8.16: Mach number, $M_\infty = 3$, $\alpha = 7°$

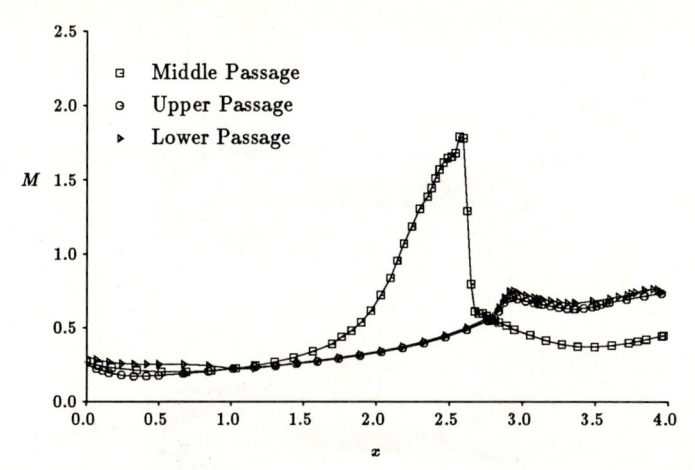

Figure 8.17: Mach number in all three passages, $M_\infty = 3, \alpha = 7°$

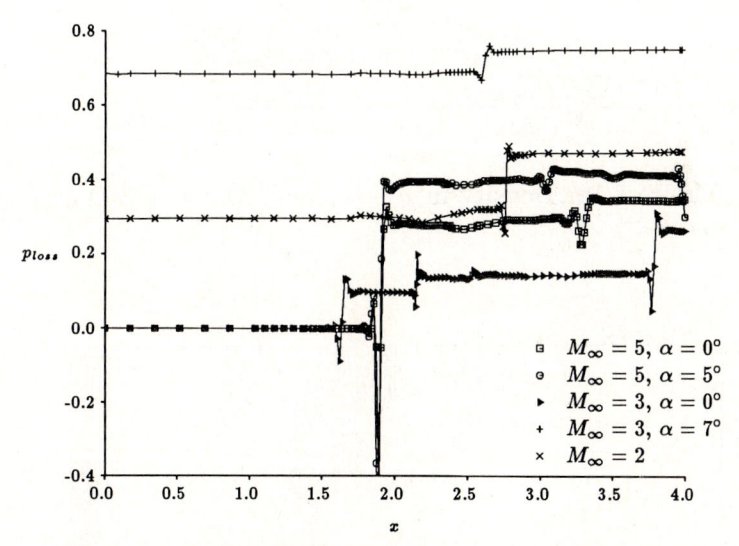

Figure 8.18: Mid-channel total pressure loss, all cases

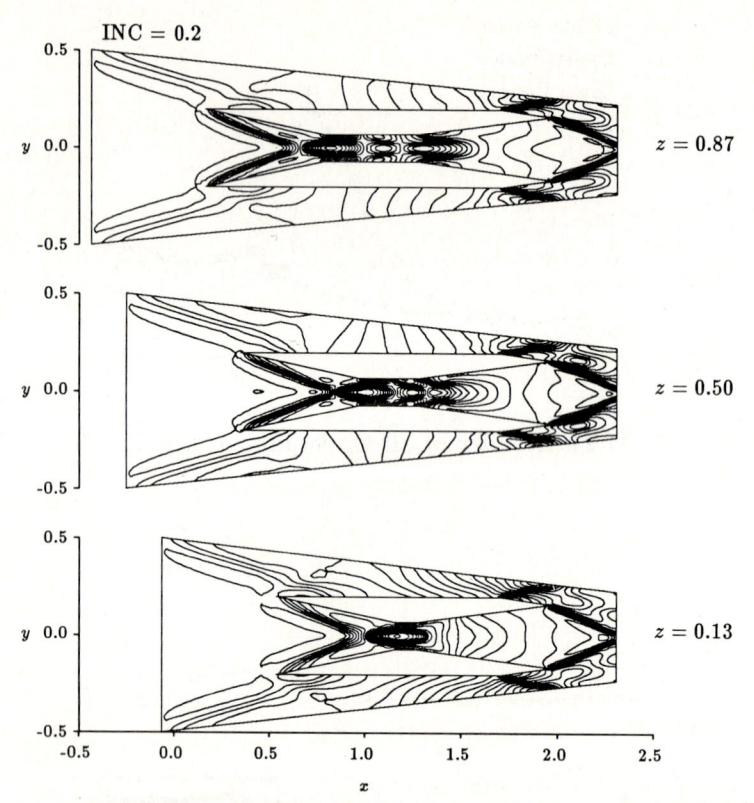

Figure 8.19: Density in $x-y$ plane, $M_\infty = 5$, 3-D case

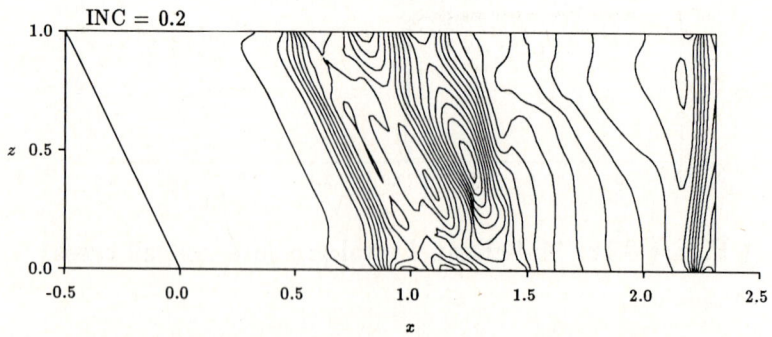

Figure 8.20: Density at $y = 0$ plane, $M_\infty = 5$, 3-D case

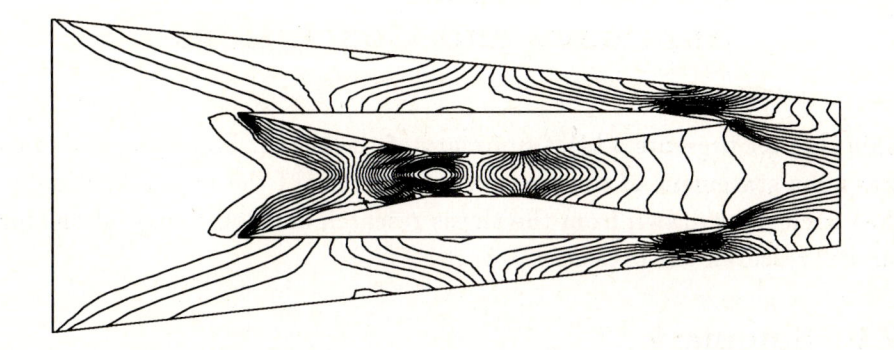

Figure 8.21: Pressure at $z = 0.5$, $M_\infty = 4$, 3-D case

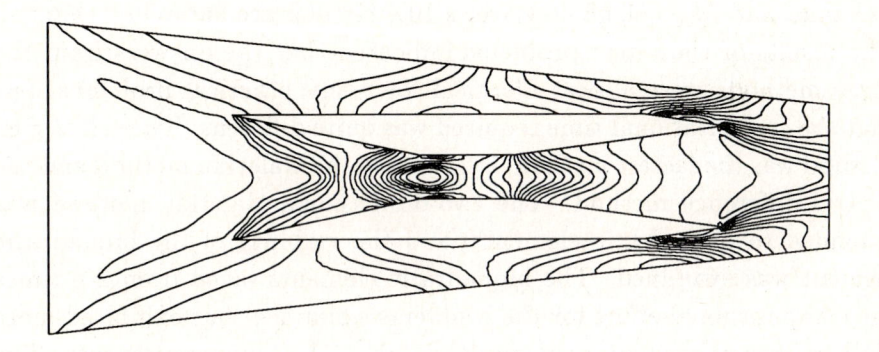

Figure 8.22: Density at $z = 0.5$, $M_\infty = 4$, 3-D case

Chapter 9
Summary and Conclusions

This chapter presents a brief summary of the thesis. Following the summary is a statement of the major contributions of the research. Finally, the conclusions drawn from the thesis research and some suggestions for further research are presented.

9.1 Summary

This thesis presented a derivation of a finite element strategy for solving the steady Euler equations. Using bilinear elements in two dimensions, the Galerkin, cell-vertex, and central difference methods were derived. The general finite element requirements for consistency and conservation were presented, and the three schemes were shown to be both consistent and conservative. The formulations were compared for $M_\infty = 2$ flow over a 5 degree ramp in a channel, $M_\infty = 4$ flow over a 15 degree ramp in a channel, $M_\infty = 1.4$ flow over a 4% circular arc bump in a channel and $M_\infty = 0.68$ flow over a 10% circular arc bump in a channel. The results of these test problems indicated that the answers from the three methods were almost identical for many practical problems, but that the computational time required was quite different. The cell-vertex method was the fastest method, followed by the Galerkin method and the central difference method. The two-dimensional Galerkin method was extended to biquadratic elements, and the viability of the biquadratic element was examined. The biquadratic elements saved from 2-6 times the computational effort for the problems examined, but exhibited more high-frequency oscillations near shocks than the bilinear elements. The cell-vertex method was extended to three dimension, and two channel flow problems were used to verify the correct implementation of the method.

Next, the thesis explored the use of adaptive grid methods in two and three dimensions. The interface formulations were presented along with a discussion on conservation for the two-dimensional interface. Two indicators for determining where to adapt were introduced, one based on a first difference of density and the other based on a second difference of density scaled by a first difference of density. Test computations in two and three dimensions verified the use of adaptation and compared the adaptation indicators. For this limited range of problems, the first difference indicator was shown to be slightly better for transonic flow, while the second difference indicator was shown to be slightly better for supersonic flow. Examples showing the ability of adaptation to reduce the sensitivity of a problem to an initially poor computational grid were presented. The reductions in computational time obtained by the use of adaptive techniques were discussed.

Chapter 7 discussed some of the dispersive errors due to the spatial discretization of the Euler equations. The concept of spatial group velocity was developed and used to predict the location of low wave number oscillations present near discontinuities in supersonic flow. The Galerkin method was shown to be fourth-order accurate on a uniform mesh, and the cell-vertex method was shown to be second-order accurate on any parallelogram mesh. Some examples demonstrated that the central difference method produced more dispersive error than the Galerkin and cell-vertex methods.

The next chapter applied the algorithms to a problem of current interest: the flow in a scramjet inlet. These examples demonstrated the ability of the computational method to model a highly complex flow at reasonable cost, and illustrated some of the interesting inviscid physics present in these inlet flows. For example, one test case illustrated that it was possible to unstart the inlet by changing the inflow angle.

9.2 Contributions of the Thesis

There are four main contributions from this thesis.

1. An adaptive finite element algorithm for quadrilateral and hexahedral elements using the multistage time integration method was developed. The algorithm was demonstrated to be robust, flexible

and accurate enough to compute a wide variety of physical problems. To this author's knowledge this work is the first application of the multistage time integration scheme to the finite element method, and the first use of adaptive hexahedral elements. The first is significant because the multistage time integration method has better stability properties than the two-step Lax-Wendroff time integration method when used with adaptive meshes [58], and the second is significant because for a given number of nodes a hexahedral mesh will have about one-fifth the elements of a tetrahedral mesh. The latter can result in significant memory and CPU savings, enabling larger problems to be computed.

2. The biquadratic finite element was introduced, and the ability of this element to save significant amounts of computational time was demonstrated. The results in this thesis show that, for some problems, savings of a factor of 6 in computational effort over bilinear elements can be obtained with biquadratic elements.

3. The Galerkin, cell-vertex, and central difference finite element methods were compared. The cell-vertex approximation is shown equivalent to a node-centered finite volume approximation, and the central difference approximation is shown equivalent to a cell-based finite volume scheme on a parallelogram mesh. The Galerkin and cell-vertex discretizations were demonstrated to be fast and robust, while the central difference discretization was shown to be inferior. The comparison also demonstrated that the terms finite element/volume/difference can often be used interchangeably for unstructured mesh algorithms.

4. The dispersive errors produced by the numerical solution schemes were analyzed. This analysis explains some of the oscillations seen near shocks in many problems, and shows that not all oscillations are due to inadequate artificial viscosity. The analysis allows one to predict the location of the oscillations, giving one a better idea of what kind of grid to use and which algorithm to choose. This analysis applies to all discrete approximations of the Euler equations, not just the three methods described in this thesis. Finally, the analysis gives insight into the order-of-accuracy issues in the various approximations, showing that the Galerkin method is

fourth order on uniform meshes and that the cell-vertex method is second-order on any parallelogram mesh, independent of grid stretching.

9.3 Conclusions

There are five major conclusions to be drawn from this research. First, the issue of structured vs. unstructured grid is more important than the finite element vs. finite volume vs. finite difference issue, since finite volume and finite difference schemes can be derived as finite element schemes. Unstructured grid methods allow much greater flexibility in problem solving, as demonstrated in chapter 8. The disadvantages of the unstructured grid methods are that they are more difficult to program, they require more memory, and they often require more computational time due to use of indirect addressing. The second disadvantage is becoming less important as computer memory sizes increase, and the third disadvantage should become less important as computer architectures are optimized for the scatter/gather operations used by the unstructured codes.

The second conclusion is that biquadratic elements provide improved accuracy at lower cost. The test problems computed showed savings of factors of 2 to 6 in CPU over the bilinear elements. While the solutions obtained tend to be more oscillatory then the bilinear solutions, for many problems these oscillations are not excessive, especially if one is interested in integral quantities such as lift, drag or moment.

The third conclusion is that, for bilinear elements, the Galerkin and cell-vertex methods are good algorithms to use, while the central difference method is not. There are three reasons for this conclusion. First, the stability limit for the central difference method is much lower than the stability limits for the other two methods, resulting in longer computational times. Second, the dispersive properties of the central difference method are much poorer. The oscillations due to dispersion persist over a much greater distance and have larger amplitude than either of the other two methods. For some problems, such spurious oscillations are undesirable, and can lead to divergence. These two problems are common to all central difference schemes, both cell-based and node-based.

Third, the central difference method is less robust than the cell-vertex and Galerkin methods. As demonstrated in chapter 5, test cases exist which converge rapidly using the cell-vertex or Galerkin method but do not converge at all using the central difference method. The author believes that part of this lack of robustness is due to the dispersive errors in the solution, since the non-convergence is altered by grid topology.

The fourth conclusion is that dispersion accounts for some of the oscillations seen in many supersonic problems. The analysis in chapter 7 provides a means of predicting the location of pre- or post-feature oscillations as a function of Mach number, grid aspect ratio, and solution method that is verified by numerical experiment. The ability to predict the location of these oscillations could be of use in the selection of a solution algorithm for a particular problem. For example, in a problem with a very strong expansion, an algorithm with post-feature oscillations may not converge due to negative overshoot after the expansion, while an algorithm with pre-feature oscillations may converge.

The final conclusion is that adaptation using quadrilateral and hexahedral elements is a viable way of reducing computational cost. The results of chapters 6 and 8 indicate that depending on the problem, savings of factors of 2 to 16 can be obtained through the use of embedded meshes, with the smaller savings in problems with more flow structure. These examples also demonstrate that the savings from adaptation increase as the number of refinement levels is increased. Adaptation still remains more of an art than a science, however, as indicated by the need for different adaptation indicators for different problems, and by the use of different thresholds for the same indicator in different problems.

9.4 Areas for Further Exploration

There are several areas which warrant further investigation. The first area is the use of biquadratic elements. This thesis has demonstrated the utility of these elements, but there are problems relating to the numerical dissipation models which need to be investigated further. Another potential research area is the use of mixed bilinear and biquadratic elements. The bilinear elements should prove superior near discontinuities, while the biquadratic elements work best in smooth portions of the flow field. A marriage of these elements might allow the best features of both elements to be obtained. The dispersion analysis presented here should be extended to include the artificial dissipation models, and should be extended to three dimensions. There is room for improvement in the selection of indicators for use in adaptation. Since usually an engineer can look at a solution and see where more resolution is needed, the application of image-processing techniques may provide new insight into the where's of adaptation. Finally, the algorithm should be extended to include viscous and chemical effects.

Appendix A
Computational Issues

A.1 Introduction

This appendix discusses some of the computational issues involved in the implementation of the finite element algorithm. The first section examines some of the methods used to implement unstructured grid algorithms on vector-parallel supercomputers. Next, the issue of computer memory requirements is addressed and ways to reduce the memory required at any particular time are proposed. Finally, a discussion of some of the data structures used in the adaptive algorithm is presented.

A.2 Vectorization and Parallelization Issues

This section discusses some of the issues involved in vectorizing and/or parallelizing the main solver routine. Since the adaptation routine is executed infrequently, no effort was made to make it vectorizable or parallelizable. In the finite element method as implemented, operations can be divided into main three classes: operations on elements, operations on nodes, and operations distributing information from elements to nodes.

A typical operation on an element involves calculating some quantity in that element based on the values of some nodal quantity. For example, the calculation of the average pressure in an element requires the pressure at the nodes of the element. This class of operations requires that the computer have a fast *gather* operation. A gather operation is one in which an array is used as an index to another array, as in the following example:

```
      DO 10 IEL = 1,NE
         TOTAL(IEL) = Q(IC(1,IEL)) + Q(IC(2,IEL))
   10 CONTINUE
```

Here, the array IC is an indexing array containing the nodes which make up element IEL. On a machine with a hardware gather operation (such as a Cray XMP, Alliant, IBM 3090, etc.) this class of operation can be computed vector and parallel.

The next class of operation, operations on nodes, occurs in such places as the flux calculation, in which quantities at a node are used to calculate other quantities at that node. These operations are the "classic" vector operations, and will vectorize on almost any machine. A subclass of this operation is a conditional node-to-node operation. For example, the following fragment from the flux calculation routine checks to make sure that the pressure is non-negative before updating the nodal pressure:

```
              PRESS = (GAMMA - 1)*(STVEC(I,5) -
     &              .5*STVEC(I,1)*(UL**2 + VL**2 + WL**2))
C
              IF (PRESS.GE.0) THEN
                  P(I) = PRESS
              ELSE
                  IPRESLIM = 1
              ENDIF
              A(I) = SQRT(GAMMA*P(I)/STVEC(I,1))
```

This type of operation vectorizes on most recent vector machines, such as the Alliant and IBM 3090.

The third class of operation, distribution of information from elements to nodes, is the most difficult operation to perform [42]. In order to avoid the possibility of element A writing something to node a before element B is finished with node a, certain precautions need to be taken. To solve this problem, the grid is preprocessed and some data structures are set up containing element-node pairs. These pairs are divided up into large groups such that each node appears only once in a group. The distribution operation is then performed on each group in succession. As currently implemented, the mesh is divided into as many groups as there are nodes per element, plus one additional group which is computed in scalar mode. For all the meshes described used in this thesis, this final scalar group contained no members, so the code was completely vectorized. The following code fragment illustrates this process:

```
      C
      C Distribute to nodes
```

```
      C
            DO 1 J = 1,NNE
              DO 2 I = ICINVP(J-1)+1,ICINVP(J)
                QNODE(ICINV1(I)) = QNODE(ICINV1(I)) + QELEM(ICINV(I))
 2          CONTINUE
 1          CONTINUE
      C
      C Exceptions
      C
      CVD$   novector   (no vector operations)
      CVD$   noconcur   (no parallel operations)
            DO I = 1,NUM_VEXCEPT
               IEL = IVEXE(I)
               INN = IVEXN(I)
               NODE = IC(INN,IEL)
               QNODE(NODE) = QNODE(NODE) + QELEM(IEL)
            ENDDO
```

where the array ICINV1 contains the node part of the node/element pairs and ICINV contains the element part of the pair. ICINVP is an array containing the index of the last pair in each pass. The scalar exceptions are handled separately, in the arrays IVEXE and IVEXN. The alternative to taking these precautions is to run this operation as a scalar operation, however, studies on the Alliant indicate that as much as 14% of the computational effort is spent in this kind of operation, so that running scalar is not an acceptable solution. This algorithm is a type of "coloring" algorithm, only the nodes are colored instead of the more usual element coloring. Other algorithms for performing this distribution exist, but their discussion is beyond the scope of this thesis.

There are a few other classes of operations, but all fall into one of the basic ideas above. For example, the calculation of an average state vector at an interface node is technically a node-to-node operation, but requires the same kinds of architectural features as the node-to-element operations above.

A.3 Computer Memory Requirements

This section discusses the memory required by the two and three dimensional adaptive finite element codes. Unstructured grid codes tend to use more memory than structured grid codes, and adaptive codes tend to use more memory than non-adaptive codes (for the same number of nodes).

One advantage of quadrilateral or hexahedral elements is that the number of elements is approximately equal to the number of nodes. In the discussions following, a *point* will refer to one element or node, depending on which is appropriate. For example, in three dimensions the element connectivity array IC has 8 entries per element, so IC will be referred to as requiring 8 words per point.

A.3.1 Two-Dimensional Memory Requirements

The two-dimensional code requires roughly 95 words of memory per point. Of these 95 words, 7 are used only if there is mesh embedding and adaptation. Also, there are several quantities that are precalculated to save computational time. If all precalculated quantities are recalculated at every iteration, an additional 14 words of memory can be saved. Finally, about 11 more words could be eliminated at the expense of code flexibility and readability. Thus, the memory required for the two dimensional code can range from about 63 words per point for a stripped-down program without embedding and adaptation, to 95 words per point for the full, flexible code. On a 24 megabyte Alliant with 16 megabytes of free memory, this means that problems as large 42,000 points can be computed. This is sufficient for most inviscid problems.

A.3.2 Three-Dimensional Memory Requirements

The three-dimensional code requires roughly 118 words of memory per point. Of these 118 words, 9 are used only if there is mesh embedding and adaptation. Recalculating all precalculated quantities would save an additional 24 words of memory. Finally, about 10 more words could be eliminated at the expense of code flexibility and readability. Thus, the memory required for the three-dimensional code can range from about 75 words per point for a stripped-down program without embedding and adaptation, to 118 words per point for the full, flexible code. Although this is more memory than required by most three-dimensional structured grid codes, it is still much lower than the 310-610 words per point mentioned by Löhner [41]. On a 24 megabyte Alliant with 16 megabytes of free memory, this means that problems as large 33,000 points can be computed. This is not sufficient for many three-dimensional problems. A typical three-dimensional problem might require as many as 250,000 points, and there are examples in the literature of problems involving well over one million points [59]. The latest generation machines from IBM and Cray, as well as many new entries into the market, can handle these large memory requirements.

As a final note on memory requirements, the memory sizes in the latest generation of supercomputer are becoming large enough so that the class of computable problems is being determined more by CPU speed than by memory limitations. This is a boon to the programmer, since it allows one to write in a clear (and hence less error-prone) style without worrying much about the memory requirements.

A.4 Data Structures for Adaptation

This section briefly describes the data structures used for mesh adaptation, and discusses a few of the key operations involved in adapting and unadapting a grid. Much of this material is common to two and three dimensions; those items that are different are discussed following the general information about the data structure.

The main arrays in the data structure are the connectivity array IC(NODE,ELEMENT), which contains the global node number of internal node NODE in element ELEMENT; the the coordinate array X(COORD,NODE), which contains the coordinates of global node NODE; and the node information array NINFO(2,NODE) which contains the nodes to the left and right of node NODE, for use in calculating normal vectors and for calculation of quantities at interface nodes. In three dimensions, NINFO has four entries per node, and contains the 4 nodes that are averaged for face interfaces or the two nodes that are averaged for edge interfaces. In three dimensions NINFO plays no part in the calculation of surface normals. These three arrays are used by the solver.

The mesh adapter adds four arrays. IFINFO(FACE,ELEM) contains the element adjacent to element ELEM on face FACE. If the face is a boundary face, IFINFO will be less than or equal to zero. If the adjacent element is at a finer level, then IFINFO contains the element number of the parent (in three dimensions) or either of the bordering elements (in two dimensions). The second array, IDEPTH(ELEM) contains the adaptation depth of the element. Each time an element is subdivided, its depth increases by 1. The third array, ICREATE(ELEM), contains the depth at which the element *first* appeared in the mesh. The fourth array, IPARENT(ELEM) contains the element from which the element was *first* created. If the element was on the base grid IPARENT is zero.

While there are many operations that are performed on the data structure during mesh adaptation, there are two which are complicated enough to warrant further explanation. These operations are determining what 4 (or 8) fine elements, called *children*, make up a coarse element, and determining which elements are adjacent to an element. The former operation is the same in two and three dimensions, while the latter operation is handled differently in two and three dimensions.

A.4.1   Finding the Children of an Element

In order to unrefine, it is necessary to find the children of an element. This requires one know the element "family", in which the children were all created at a certain level. The temporary array ICHILD stores the children for a particular element. It is generated in the following manner:

1. Search through all the elements, looking for elements with ICREATE equal to the desired child depth.

2. When an element is found, get its parent from IPARENT.

3. Check ICHILD(1,PARENT). This contains the number of children of the element found so far. Put the child element number in the ICHILD array.

4. When all elements have been searched, go back and check the first element of ICHILD, and if it is non-zero, set ICHILD(1,ELEM) = ELEM. The process is complete, and ICHILD will contain the 4 (or 8) elements that resulted from a division.

This process requires one complete sweep through all the elements, so it is only practical to generate this array once or twice per level. This is not a problem since the adaptation algorithm only works with refining and unrefining elements of one depth at a time. In the following discussions, the term "ICHILD generated for a level" means that the children are of the desired level.

A.4.2 Finding The Adjacent Element

When an element is subdivided, it is necessary to find the elements adjacent to all the faces in order to keep the face pointer array IFINFO correctly updated. There are three cases: the adjacent element is at a lower (coarser) level, the adjacent element is at the same level, or the adjacent element is at a higher (finer) level. The first two cases are treated identically in two and three dimensions, and are trivial. In both of these cases, the adjacent element (there is only one) is given directly in IFINFO. The difficult case is the case in which the adjacent elements are finer.

Three Dimensions

In three dimensions, when one looks for the elements on a particular face, one needs to know where on that face the elements are (which element on the face shares node i on that face). As implemented, this search requires that ICHILD be generated for the finer of the two levels. In three dimensions, IFINFO points to the parent of the adjacent elements. The following procedure finds the 4 elements on the face FACE of element IEL:

1. Search through the 8 child elements to find the four with their IFINFO pointing back at IEL.

2. Once these elements have been found, search through their IC arrays to find which elements share which nodes on the face of IEL.

3. Return the 4 element numbers.

Two Dimensions

In two dimensions without directional embedding, the same method as used in three dimensions can be used. As the code currently stands, the hooks for directional embedding are still present, so a more complicated search procedure is used. Since diretional embedding is no longer used, the procedure will not be described here.

Appendix B
Scramjet Geometry Definition

This appendix presents the exact geometry used for the scramjet calculations of chapter 8. Figure B.1 gives the geometry of the 2-D scramjet. Figure B.2 gives an $x - y$ slice of the geometry of the 3-D scramjet at $z = 0$, Fig. B.3 gives an $x - y$ slice of the geometry of the 3-D scramjet at $z = 1$.

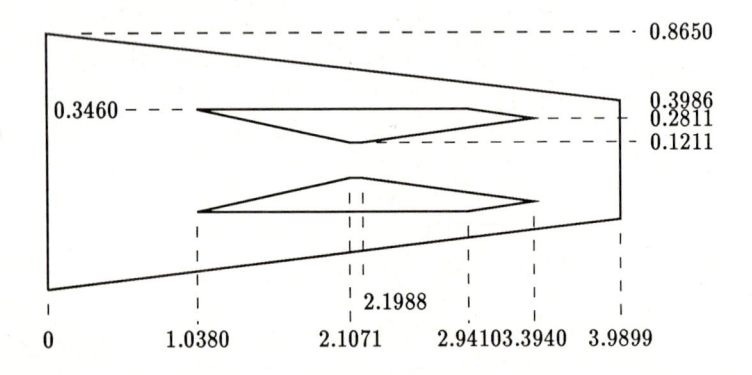

Figure B.1: Geometry of Two-Dimensional Scramjet

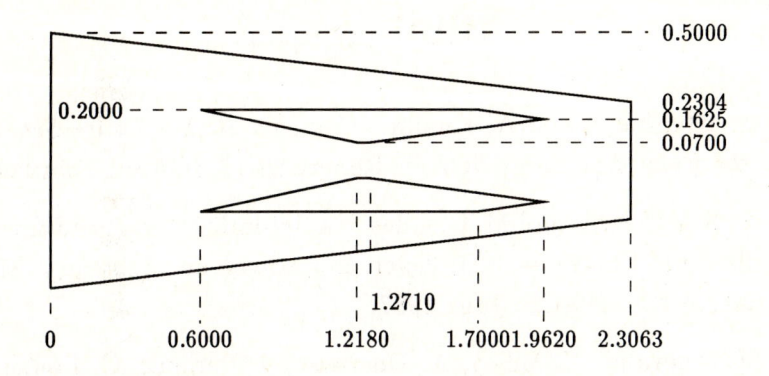

Figure B.2: Geometry of Three-Dimensional Scramjet at $z = 0$

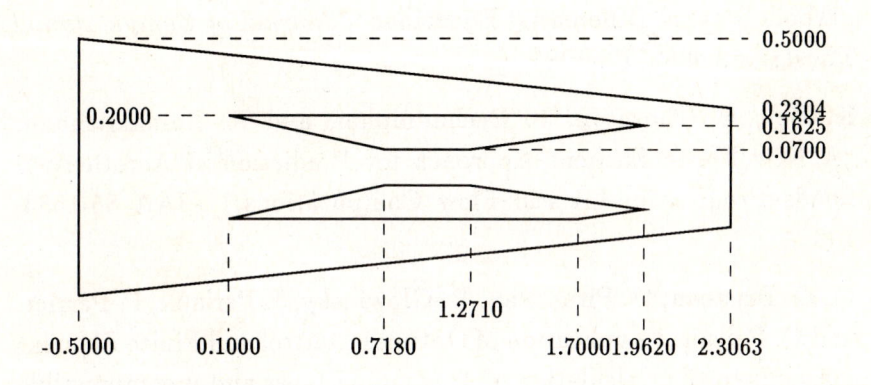

Figure B.3: Geometry of Three-Dimensional Scramjet at $z = 1$

Bibliography

[1] S. Abarbanel and A. Kumar. *Compact Higher-Order Schemes for the Euler Equations.* ICASE Report 88-13, ICASE, February 1988.

[2] S. R. Allmaras and M. B. Giles. "A Second Order Flux Split Scheme for the Unsteady 2-D Euler Equations on Arbitrary Meshes." AIAA 87-1119-CP, June 1987.

[3] F. Angrand, V. Billey, A. Dervieux, J. Periaux, C. Pouletty, and B. Stoufflet. "2-D and 3-D Euler Flow Calculations with a Second-Order Accurate Galerkin Finite Element Method." AIAA 85-1706, July 1985.

[4] M. J. Berger. *Adaptive Finite Difference Methods in Fluid Dynamics.* Technical Report DOE/ER/03077-277, Courant Mathematics and Computing Laboratory, New York University, September 1984.

[5] M. J. Berger and J. Oliger. "Adaptive Mesh Refinement for Hyperbolic Partial Differential Equations." *Journal of Computational Physics*, 53:484–512, 1984.

[6] K. Bey, E. Thornton, P. Dechaumphai, and R. Ramakrishnan. "A New Finite Element Approach for Prediction of Aerothermal Loads–Progress in Inviscid Flow Computations." AIAA 85-1533, 1985.

[7] M. O. Bristeau, O. Pironneau, R. Glowinsky, J. Periaux, P. Perrier, and G. Poirier. "Application of Optimal Control and Finite Element Methods to the Calculation of Transonic FLows and Incompressible Viscous Flows." In *Proc. IMA Conference on Numerical Methods in Applied Fluid Mechanics*, Academic Press, 1980.

[8] F. E. Dabaghi, J. Periaux, O. Pironneau, and G. Poirier. "Three-Dimensional Finite Element Solution of Steady Euler Transonic Flow by Stream Vector Correction." AIAA 85-1532, 1985.

[9] R. L. Davis. *The Prediction of Compressible, Viscous Secondary Flow in Channel Passages*. PhD thesis, University of Connecticut, 1982.

[10] J. Donéa. "A Taylor-Galerkin Method for Convective Transport Problems." *Int. Journal for Numerical Methods in Engineering*, 20:101–119, 1984.

[11] A. Ecer and H. U. Akay. "Finite Element Formulation of the Euler Equations for the Solution of Steady Transonic Flows." *AIAA Journal*, 21:410–416, March 1983.

[12] A. Ecer, J. Spyropoulos, and M. Sims. "Parallel Processing Techniques for the Solution of the Euler Equations." AIAA 88-0620, 1988.

[13] P. R. Eiseman. "Adaptive Grid Generation." *Computer Methods in Applied Mechanics and Engineering*, 64:321–376, 1987.

[14] A. P. French. *Vibrations and Waves*. W. W. Norton and Company, 1971.

[15] M. M. Hafez, W. G. Habashi, S. M. Przybytkowski, and M. F. Peeters. "Compressible Viscous Internal Flow Calculations by a Finite Element Method." AIAA 87-0644, 1987.

[16] M. G. Hall. *Cell Vertex Schemes for the Solution of the Euler Equations*. Technical Memo Aero 2029, Royal Aircraft Establishment, March 1985.

[17] T. Hughes, L. Franca, I. Harari, M. Mallet, F. Shakib, and T. Spelce. "Finite Element Method for High-Speed Flows: Consistent Calculation of Boundary Flux." AIAA 87-0556, 1987.

[18] T. Hughes, L. Franca, and M. Mallet. "A New Finite Element Formulation for Computational Fluid Dynamics: I. Symmetric Forms of the Compressible Euler and Navier-Stokes Equations and the Second Law of Thermodynamics." *Computer Methods in Applied Mechanics and Engineering*, 54:223–234, 1986.

[19] T. Hughes and M. Mallet. "A New Finite Element Formulation for CFD: The Generalized Streamline Operator for Multidimensional Advective-Diffusive Systems.". 1986. *Computer Methods in Applied Mechanics and Engineering.*

[20] J. F. D. III. *Grid Adaptation for Complex Two-Dimensional Transonic Flows.* PhD thesis, M.I.T., August 1987.

[21] J. F. D. III and J. R. Baron. "Grid Adaptation for the 2-D Euler Equations." AIAA 85-0484, January 1985.

[22] J. F. D. III and J. R. Baron. "Robust Grid Adaptation for Complex Transonic Flows." AIAA 86-0495, January 1986.

[23] B. Irons. "The Superpatch Theorem and Other Propositions Relating to the Patch Test." In *Proceedings of the Fifth Canadian Congress of Applied Mechanics*, pp. 651–652, 1975.

[24] A. Jameson. "The Evolution of Computational Methods in Aerodynamics." *ASME Journal of Applied Mechanics*, 50:1052–1069, 1983.

[25] A. Jameson. "Successes and Challenges in Computational Aerodynamics." AIAA 87-1184, 1987.

[26] A. Jameson. "A Vertex Based Multigrid Algorithm for Three Dimensional Compressible Flow Calculations." In T. E. Tezduyar and T. J. R. Hughes, editors, *Numerical Methods for Compressible Flow: Finite Difference, Element and Volume Techniques*, American Society of Mechanical Engineers, New York, NY, 1986.

[27] A. Jameson, T. J. Baker, and N. P. Weatherhill. "Calculation of Inviscid Transonic Flow over a Complete Aircraft." AIAA 86-0103, 1986.

[28] A. Jameson, W. Schmidt, and E. Turkel. "Numerical Solutions of the Euler Equations by a Finite Volume Method Using Runge-Kutta Time Stepping Schemes." AIAA 81-1259, June 1981.

[29] J. Kallinderis and J. R. Baron. "Adaptation Methods for a New Navier-Stokes Algorithm." AIAA 87-1167, June 1987.

[30] S. Kennon. "An Optimal Spatial Accuracy Finite Element Scheme for the Euler Equations." AIAA 88-0035, January 1988.

[31] A. Kumar. "Numerical Analysis of the Scramjet-Inlet Flow Field by Using Two-Dimensional Navier-Stokes Equations." NASA Technical Paper 1940, December 1981.

[32] A. Kumar. "Numerical Simulation of Flow Through a Two-Strut Scramjet Inlet." AIAA 85-1664, 1985.

[33] A. Kumar. "Numerical Simulation of Scramjet Inlet Flow Fields." NASA Technical Paper 2517, 1986.

[34] P. Kutler. "Supersonic Flow in the Corner by Two Intersecting Wedges." *AIAA Journal*, 12:577–578, 1974.

[35] P. Lax. "Weak Solutions of Nonlinear Hyperbolic Equations and Their Numerical Computation." *Communications of Pure and Applied Mathematics*, 7:159–193, 1954.

[36] H. Liepmann and A. Roshko. *Elements of Gasdynamics*. John Wiley and Sons, New York, 1957.

[37] D. Lindquist and M. Giles. "A Comparison of Numerical Schemes on Triangular and Quadrilateral Meshes.". June 1988. Submitted to 11th International Conference on Numerical Methods in Fluid Dynamics.

[38] D. R. Lindquist. *A Comparison of Numerical Schemes on Triangular and Quadrilateral Meshes*. Master's thesis, M.I.T., May 1988.

[39] R. Löhner. "The Efficient Simulation of Strongly Unsteady Flows by the Finite Element Method." AIAA 87-0555, January 1987.

[40] R. Löhner. "FEM-FCT and Adaptive Refinement Schemes for Strongly Unsteady Flows." In *Proc. ASME Winter Annual Meeting*, Anaheim, California, December 1986.

[41] R. Löhner. "Finite Elements in CFD: What Lies Ahead.". Presented at the World Conference on Computational Mechanics, Austin, Texas, September, 1986.

[42] R. Löhner and K. Morgan. "Finite Element Methods on Supercomputers: The Scatter Problem." In *Proceedings of the NUMETA '85 Conference, Swansea*, pp. 987–990, 1985.

[43] R. Löhner, K. Morgan, J. Peraire, and O. C. Zienkiewicz. "Finite Element Methods for High Speed Flows." AIAA 85-1531, 1985.

[44] R. Löhner, K. Morgan, and O. C. Zienkiewicz. "The Solution of Non-linear Hyperbolic Equation Systems by the Finite Element Method." *Int. Journal for Numerical Methods in Fluids*, 4:1043–1063, 1984.

[45] R. W. MacCormack. "The Effect of Viscosity in Hypervelocity Impact Cratering." AIAA 69-354, May 1969.

[46] R. Magnus and H. Yoshihara. "Inviscid Transonic Flow over Airfoils." *AIAA Journal*, 8:2157–2162, 1970.

[47] D. Mavriplis. *Solution of the Two-Dimensional Euler Equations on Unstructred Triangular Meshes*. PhD thesis, Princeton University, June 1987.

[48] K. Morgan, R. Löhner, J. R. Jones, J. Peraire, and M. Vahdati. *Finite-Element FCT for the Euler and Navier-Stokes Equations*. Technical Report C/R/540/86, Institute for Numerical Methods in Engineering, February 1986.

[49] J. Murphy. "Application of the Generalized Galerkin Method to the Computation of Fluid Flows." In *Proceedings of the First AIAA CFD Conference*, pp. 63–68, 1973.

[50] K. Nakahashi and S. Obayashi. "Viscous Flow Computations Using a Composite Grid." AIAA 87-1128, 1987.

[51] R. H. Ni. "A Multiple-Grid Scheme for Solving the Euler Equations." *AIAA Journal*, 20(11):1565–1571, November 1982.

[52] J. Oden, T. Strouboulis, P. Devloo, L. W. Spradley, and J. Price. "An Adaptive Finite Element Strategy for Complex Flow Problems." AIAA 87-0557, January 1987.

[53] J. Peraire, K. Morgan, J. Peiro, and O. C. Zienkiewicz. "An Adaptive Finite Element Method for High Speed Flows." AIAA 87-0558, January 1987.

[54] J. Peraire, M. Vahdati, K. Morgan, and O. C. Zienkiewicz. "Adaptive Remeshing for Compressible Flow Computations." *Journal of Computational Physics*, 72:449–466, October 1987.

[55] E. Perez. *Finite Element and Multigrid Solution of the Two Dimensional Euler Equations on a Non-structured Mesh*. Rapport de Recherche 442, INRIA, September 1985.

[56] K. Powell. *Vortical Solutions of the Conical Euler Equations*. PhD thesis, M.I.T., July 1987.

[57] R. J. Prozan, L. W. Spradley, P. G. Anderson, and M. L. Pearson. "The General Interpolants Method: A Procedure for Generating Numerical Analogs of the Conservation Laws." In *Proceedings of the AIAA Third Computational Fluid Dynamics Conference*, pp. 106–115, 1977.

[58] R. Ramakrishnan, K. S. Bey, and E. A. Thornton. "An Adaptive Quadrilateral and Triangular Finite Element Scheme for Compressible Flows." AIAA 88-0033, January 1988.

[59] A. Rizzi. "Three-Dimensional Solutions to the Euler Equations with One Million Grid Points." *AIAA Journal*, 23(12), December 1985.

[60] A. Rizzi and L. E. Eriksson. "Computation of Flow Around Wings Based on the Euler Equations." *Journal of Fluid Mechanics*, 148, November 1984.

[61] P. L. Roe. *Error Estimates for Cell-Vertex Solutions of the Compressible Euler Equations*. ICASE Report 87-6, ICASE, January 1987.

[62] P. E. Rubbert and E. N. Tinoco. "Impact of Computational Methods on Aircraft Design." AIAA 83-2060, August 1983.

[63] R. Shapiro and E. Murman. "Adaptive Finite Element Methods for the Euler Equations." AIAA 88-0034, January 1988.

[64] R. Shapiro and E. Murman. "Cartesian Grid Finite Element Solutions to the Euler Equations." AIAA 87-0559, January 1987.

[65] R. A. Shapiro. *Analysis of Distortion in Assumed Displacement Finite Elements.* Master's thesis, M.I.T., May 1984.

[66] S. Shirayama and K. Kuwahara. "A Zonal Approach for Computation of Unsteady Incompressible Viscous Flow." AIAA 87-1140, 1987.

[67] B. Stoufflet, J. Periaux, F. Fezoui, and A. Dervieux. "Numerical Simulation of 3-D Hypersonic Euler Flows Around Space Vehicles Using Adapted Finite Elements." AIAA 87-0560, January 1987.

[68] G. Strang. *Introduction to Applied Mathematics.* Wellesley-Cambridge Press, Wellesley, Massachusetts, 1986.

[69] G. Strang and G. Fix. *An Analysis of the Finite Element Method.* Prentice-Hall, Inc., Englewood Cliffs, NJ, 1973.

[70] J. F. Thompson. "A Survey of Dynamically-Adaptive Grids in Numerical Solution of Partial Differential Equations." AIAA 84-1606, June 1984.

[71] E. Thornton, R. Ramamkrishnan, and P. Dechaumphai. "A Finite Element Approach for Solution of the 3D Euler Equations." AIAA 86-0106, 1987.

[72] W. J. Usab. *Embedded Mesh Solutions of the Euler Equations Using a Multiple-Grid Method.* PhD thesis, M.I.T., December 1983.

[73] W. J. Usab and E. M. Murman. *Embedded Mesh Solutions of the Euler Equations Using a Multiple-Grid Method*, pp. 447–472. Pineridge Press, 1985.

[74] R. Vichnevetsky and J. B. Bowles. *Fourier Analysis of Numerical Approximations of Hyperbolic Equations.* SIAM, 1982.

[75] M. E. White, J. P. Drummond, and A. Kumar. "Evolution and Application of CFD Techniques for Scramjet Engine Analysis." *AIAA Journal of Propulsion*, 3(5):423–439, September 1987.

List of Symbols

$()^{(e)}$	Elemental quantity
$()_\infty$	Free stream quantity
A, B	Flux jacobians
A	Area
a	Speed of sound
$\mathcal{R}$	Aspect ratio
C_i	Riemann invariants
$\mathcal{D}_2$	Second difference operator
$\mathcal{D}_x, \mathcal{D}_y$	Derivative operator
e	Total energy
$\mathbf{F}, \mathbf{G}, \mathbf{H}$	Flux vectors
F, G, H	Flux vectors
h_0	Total enthalpy
M	Mach number
M	Consistent mass matrix
M_L	Lumped mass matrix
N	Local shape function
$\mathbf{N}$	Global shape function
$\tilde{N}$	Local test function
$\tilde{\mathbf{N}}$	Global test function
n_x, n_y	Components of unit normal
$\hat{n}$	Unit normal vector
p	Pressure
p_0	Total pressure
R	Residual vector
s_x, s_y	Difference operator Fourier stencils
Δt	Time step
$\mathbf{U}$	State vector
u, v, w	Cartesian velocities

u_n, u_t	Normal and tangential velocities
V	Artificial viscosity
Δx	Element size
x, y, z	Cartesian coordinate directions
γ	Ratio of specific heats
ρ	Density
ϕ, θ	x, y wave numbers
ν_1, ν_2	Artificial viscosity coefficients
ξ, η, ζ	Natural coordinate directions

Index

Adaptation
    conservation at interfaces, 84
    data structures used, 148
    embedded interfaces, 84–85, 87
Adaptation, CPU comparisons, 89, 91–92
Adaptation criteria, 80
    directional embedding, 83
    first-difference indicator, 81
    second-difference indicator, 82
Adaptation examples, 88
Adaptation, overview, 76–77
Adaptation procedure, 77–78
    how much adaptation?, 80
    placement of boundary nodes, 79
Artificial viscosity, see Smoothing

Bilinear element, 17
    comparison with biquadratic, 51
    geometry of, 17
    interpolation functions, 17
    jacobian matrix, 18
Biquadratic element, 19
    comparison with bilinear, 51
    geometry of, 17
    interpolation functions, 19
    spatial discretization, 27
    test cases, 51–52
    smoothing, 38

Cell-vertex method, 27
    comparison with finite-volume method, 28
    consistent mass matrix, 28
    dispersive properties, 105, 110
    residual matrix, 27
    test functions, 27
    three dimensions, 28
Central difference method, 29
    comparison with finite volume method, 29
    consistent mass matrix, 30
    dispersive properties, 110
    lumped mass matrix, 31

    residual matrix, 29
    test functions, 29
Characteristic variables, 9, 32
10% Circular Arc Bump, 46–47
    adaptive solution, 89–90
    computed solution, 52
Conservation, 39
    numerical verification, 48
    preservation at embedded interfaces, 84
    proof of, 39
    smoothing, 40
Conservation, proof of, 40
Consistent mass matrix, 24
    cell-vertex method, 28
    central difference method, 30
    galerkin method, 26
15° Converging Channel, 43, 45
    adaptive solution, 88–89
    computed solution, 44, 46, 51–52
    exact solution, 44–45
    three dimensions, 53
10% Cosine Bump, 47, 52
CPU comparisons
    adaptive grids, 89, 91–92
    base algorithms, 48
    biquadratic vs. bilinear elements, 51, 123

Degenerate elements, 13
Dispersion
    dispersion relation, 107–108
    Fourier analysis, 107
    group velocity, 108
    numerical verification of, 111–112
    prediction of oscillations induced by, 110
    properties of the numerical methods, 109–110
10° Double wedge, 54
    adaptive solution, 90
    geometry, 53

Embedded grids, see Adaptation, 76

Entropy, 8
    linearization of, 106
Euler equations, 5
    auxilliary quantities, 8
    boundary conditions, 8
    nondimensionalization, 7
    statement of, 6
    underlying assumptions, 5–6

Finite Element Methods
    survey, 3
Finite element terms defined, 11

Galerkin method, 26
    consistent mass matrix, 26
    dispersive properties, 104, 109
    lumped mass matrix, 27
    residual matrix, 27
Grid embedding, see Adaptation, 76
Group velocity, 108

Hexahedral elements vs. tetrahedral elements, 12

Interfaces treatments with adaptation, 84–85, 87
Interpolation functions, 14
    bilinear element, 17
    biquadratic element, 19
    requirements for, 14–15
    trilinear Element, 19
Isoparametric element, 16

Jacobian matrix, 15
    bilinear element, 18
    trilinear element, 21

Local time stepping, 38
Lumped mass matrix, 25
    central difference method, 31
    galerkin method, 27

Mach number, 8
Memory requirements, 146

Natural coordinates, 13
    calculation of derivatives in, 15
Normal vector calculation, 31
Numerical boundary conditions, 31
    open boundary, 32
    solid surface, 31

Parallelization issues, 144–146
Physical boundary conditions
    far field, 9
    solid surface, 9

Quadrilateral elements vs. triangular elements, 12

Research contributions, 139
Research Goals, 1
Residual matrix
    cell-vertex method, 27
    central difference method, 29
    galerkin method, 27
Riemann invariants, 9, 32

Scramjet Inlets
    geometry of, 120–121, 126, 152–153
    performance and total pressure loss, 125
    three dimensions, 126
    two dimensional cases, 121–125
Shape functions, see interpolation functions
Smoothing
    biquadratic elements, 38
    conservation, 40
    high accuracy second difference, 35
    low accuracy second difference, 34
    numerical effects, 49
    overall computation, 37
    pressure switch, 36
Solution algorithm
    convergence criterion, 22
    overview, 22
Spatial discretization, 23
    conservation, 39
    consistency, 39
Specification of exit pressure, 33
Speed of sound, 8
State equation, 6
Subparametric element, 16
Surface normal vector calculation, 31

Test cases, summary of, 42, 54

Test cases, three dimensions
    adaptative grids, 90
    5° converging channel, 54
    10° double wedge, 54, 90
Test cases, two dimensions
    adaptative grids, 88–91
    4% circular arc bump, 46–47, 52, 89–90
    5° converging channel, 43–46, 51, 88–89
    10% cosine bump, 47, 52
    distorted grid, 91
Test functions, 23
    cell-vertex method, 27
    central difference method, 29
    galerkin method, 26

Time integration, 38
    stability, 38
    time step computation, 38
Total enthalpy, definition, 6
Total pressure, 8
Total pressure loss, 8, 125
Trial functions, see interpolation functions
Trilinear Element, 19
    geometry of, 20
    interpolation functions, 19
    jacobian matrix, 21
    spatial discretization, 28

Vectorization issues, 144–146

Addresses of the editors of the series "Notes on Numerical Fluid Mechanics":

Prof. Dr. Ernst Heinrich Hirschel (General Editor)
Herzog-Heinrich-Weg 6
D-8011 Zorneding
Federal Republic of Germany

Prof. Dr. Kozo Fujii
High-Speed Aerodynamics Div.
The ISAS
Yoshinodai 3-1-1, Sagamihara
Kanagawa 229
Japan

Prof. Dr. Bram van Leer
Department of Aerospace Engineering
The University of Michigan
Ann Arbor, MI 48109-2140
USA

Prof. Dr. Keith William Morton
Oxford University Computing Laboratory
Numerical Analysis Group
8-11 Keble Road
Oxford OX1 3QD
Great Britain

Prof. Dr. Maurizio Pandolfi
Dipartimento di Ingegneria Aeronautica e Spaziale
Politecnico di Torino
Corso Duca Degli Abruzzi, 24
I-10129 Torino
Italy

Prof. Dr. Arthur Rizzi
FFA Stockholm
Box 11021
S-16111 Bromma 11
Sweden

Dr. Bernard Roux
Institut de Mécanique des Fluides
Laboratoire Associè au C.R.N.S. LA 03
1, Rue Honnorat
F-13003 Marseille
France